National-Scale Dynamic Water Resources Assessment Model in China

This detailed book introduces China's national dynamic water resources assessment model, presenting its construction, application scenarios, and modern modeling approach to replace traditional statistics-based methods.

This book thoroughly explores the applications of distributed hydrological modeling techniques in national water resources assessment. It presents the successful development of the China Water Assessment Model (CWAM), which is based on the WEP-L hydrological model. CWAM demonstrates broad potential in supporting national water information acquisition, uniquely covering both surveyed and un-surveyed regions. The work highlights the model's ability to account for China's diverse climatic and geological characteristics, while improving the efficiency of water resources assessment and providing solutions to dynamic assessment challenges, especially under climate change and human impacts.

This work will serve as an essential reference for scholars and students in hydrology and water resources, as well as policy makers and engineers involved in water resources management and assessment.

Huan Liu is a Senior Engineer at the China Institute of Water Resources and Hydropower Research, mainly engaged in water cycle simulation and water resources assessment.

Yangwen Jia is a Senior Engineer at the China Institute of Water Resources and Hydropower Research, mainly engaged in water cycle simulation.

Jianhua Wang is a Senior Engineer at the China Institute of Water Resources and Hydropower Research, mainly engaged in water resources assessment and utilization.

Junkai Du is a Senior Engineer at the China Institute of Water Resources and Hydropower Research, mainly engaged in water information monitoring and application.

Cunwen Niu is a Senior Engineer at the China Institute of Water Resources and Hydropower Research, mainly engaged in water resources management.

Peng Hu is a Senior Engineer at the China Institute of Water Resources and Hydropower Research, mainly engaged in protection and restoration of rivers and lakes.

Jiajia Liu is a Senior Engineer at the China Institute of Water Resources and Hydropower Research, mainly engaged in hydrological model development.

National-Scale Dynamic Water Resources Assessment Model in China

Huan Liu, Yangwen Jia, Jianhua Wang, Junkai Du,
Cunwen Niu, Peng Hu, and Jiajia Liu

CRC Press
Taylor & Francis Group
Boca Raton   London   New York

CRC Press is an imprint of the
Taylor & Francis Group, an **informa** business

Cover image created by Huan Liu.

This book is published with financial support from the National Natural Science Foundation of China (Grant No. 52394233), the National Key Research and Development Plan (Grant No. 2021YFC3200203), and the Young Elite Scientists Sponsorship Program by CAST (Grant No. 2023QNRC001).

First edition published 2026
by CRC Press

2385 NW Executive Center Drive, Suite 320, Boca Raton FL 33431

and by CRC Press

4 Park Square, Milton Park, Abingdon, Oxon, OX14 4RN

CRC Press is an imprint of Taylor & Francis Group, LLC

ISBN: 978-1-041-08697-0 (hbk)
ISBN: 978-1-041-08702-1 (pbk)
ISBN: 978-1-003-64664-8 (ebk)

DOI: 10.1201/9781003646648

Typeset in Minion
by KnowledgeWorks Global Ltd.

Contents

List of Figures

List of Tables

Preface

HYDROLOGICAL ANALYSIS AND WATER resources assessment involve a comprehensive evaluation of the quantity, quality, spatial and temporal distribution, and development potential of water resources within a river basin or region. This process is essential for planning, developing, utilizing, protecting, and managing water resources, with its outcomes serving as a critical foundation for water-related activities and decision-making. Since the mid twentieth century, numerous countries have faced challenges such as water shortages, ecological degradation of water systems, and water pollution. As a result, water resources assessment has progressively gained recognition as a vital component of water resource planning and management. In 1968 and 1978, the United States conducted two comprehensive national water resources assessments. The first assessment focused on evaluating the background state of natural water resources, while the second assessment concentrated on assessing the development and utilization of water resources. At this juncture, the methodologies and techniques for assessing water resources, primarily based on statistical analysis, were initially established. Subsequently, in 1975, several countries and regions, including Western Europe, Japan, and India, published their respective water resources assessment results. Compared to countries like the United States, China's efforts in water resources assessment began at a relatively later stage. The inaugural national water resources assessment in China was conducted in 1980. This assessment primarily adopted the methodology proposed by the United States, yet it was significantly adapted to align with China's specific circumstances. Notably, it introduced the concept of non-duplicated groundwater resources along with a tailored assessment methodology. The country is currently undertaking its third national water resources survey and assessment, aimed at elucidating the characteristics of water resource evolution under the synergistic impacts of human activities and climate change.

Intensive human activities significantly alter the hydrological cycle and the distribution of water resources in numerous river basins. A dynamic assessment of water resources, that is, the evaluation and prediction of water resource conditions and their variations within a river basin over past, present, and future timeframes, is essential for sustainable water resource development and management. However, the traditional water resources assessment methodology is grounded in the concept of "measurement-reduction-correction" of runoff. The fundamental principle of this method is to restore the measured water resources to their natural amount by eliminating anthropogenic influences through a systematic process. To address the impact of human activities on water consumption, we first quantify and aggregate the volume of water used by various human activities along with corresponding measured values. Second, a "consistency" correction method is applied to mitigate alterations in precipitation-runoff relationships caused by changes in subsurface conditions due to human activities. However, when confronted with the contemporary demands for real-time and intelligent water resources management, traditional methodologies exhibit certain limitations. These limitations specifically encompass three key areas:

i) Narrow scope of assessment: Surface water resources and groundwater resources are the objects of concern in traditional water resources assessment. Some countries even focus only on river runoff while neglecting groundwater resources. In recent years, the introduction of the concepts of blue water and green water has broadened the scope of water resources assessment.

ii) Static assessment: This methodology relies on statistical analysis and necessitates substantial human resources and time to evaluate historical changes in water resources over a defined period. As a result, national-scale water resource assessments often require several years or even a decade to complete, leading to significant inefficiencies. Moreover, this method fails to adequately capture the dynamic influences of human activities and climate change on water resources and is unable to project future changes of water resources under different scenarios.

iii) Lumped assessment of spatial and temporal scales: The traditional methodology is characterized by a lumped assessment of spatial and temporal scales. It ignores the spatial heterogeneity of hydrological parameters and natural geological conditions within the assessed river basin, resulting in outcomes that merely represent regional averages. Such results are not conducive to effective water resources management in small-scale regional units. Additionally, this methodology typically employs annual time units, ignoring the intra-annual variations in water resources, particularly during flood and drought periods. The primary reason for the limitations described above lies in lack of appropriate models that can accurately simulate water movement within a river basin. Many statistical or lumped methodologies, such as the precipitation – runoff relationship curve approach and the runoff coefficient model, are inadequate for addressing these challenges. This inadequacy stems from their lack of foundation in physical hydrological processes, which hinders their ability to reflect the impacts of land cover changes and water utilization.

Dynamic assessment of water resources has become essential for accurately reflecting the variations in water resources under the significant impacts of human activities and climate change. Therefore, several scholars attempt to develop distributed hydrological modeling techniques for water resources assessment, which has become an important tool for water resources assessment. Since the early twentieth century, Wang Hao's team has proposed water resources assessment theory and methodology based on the "natural-artificial" dualistic water cycle framework. The WEP-L (Water and Energy transfer Processes in Large river basin) model, a physically based, spatially distributed hydrological model, was developed and applied for the dynamic assessment of water resources in large river basins, including the Yellow River Basin and the Haihe River Basin. However, there is a lack of a national-scale model in China that adequately addresses the practical requirements for nationwide water resources planning and management. The expansion of detailed hydrological modeling from the catchment scale to national or continental scales holds significant theoretical and practical importance, yet it encounters numerous challenges. Drawing upon the valuable insights and experience accumulated from previous studies, a national-scale dynamic water resources assessment model in China (CWAM) has been successfully developed, building on the foundation of the WEP-L model. In CWAM, various climatic and hydrological conditions, geological structures, and their respective impacts on infiltration and runoff were systematically examined and integrated into the model. The study area, spanning 9.6 million km^2 was divided into 19,406 sub-basins and 81,687 contour belts. The computation units, model inputs, and model structure and parameters in CWAM were improved to obtain a well-defined simulation area, a more reliable input as well as a detailed description of soil moisture movement in several special vadose zones such as karst development, swelling, and frozen soil. Continuous simulations of diverse natural hydrological processes were conducted for 62 years from 1956 to 2017. The efficacy of the model was demonstrated through a comparison of simulated and statistical monthly streamflow at 203 hydrological stations across the country. For the validation period from 1981 to 2000, the Nash-Sutcliff Efficiency (NSE) exceeded 0.7 at 80% of the stations, and the absolute value of relative error (RE) was below 10% at 95% of the stations. The result highlights the benefit of incorporating new mechanisms on the special vadose zone water movement and accounts for the impact of elevation change on meteorological and vegetation variables. This technology can serve as a valuable reference for large-scale hydrological simulation with diverse climatic, topographic, and surface conditions.

Through this verified model, the study accomplished three key applications: ①The spatiotemporal patterns of critical water fluxes across China, specifically precipitation (P), runoff (R), infiltration (Inf), and actual evapotranspiration (ET_a), were illustrated. The 62-year average values of P, R, Inf, and ET_a were 678.1 mm, 275.5 mm, 322.6 mm, and 431.6 mm, respectively. Over the 62-year period, P and R exhibited decreasing trends, while Inf and ET_a showed weak increasing trends. Changes in R and Inf were primarily driven by variations in P, with correlation coefficients of 0.74 and 0.73, respectively. The ET_a was constrained by a combination of P and energy. ②Changes in water resources were assessed both nationally and across distinct water resource regions for the period 1956–2017. The national multi-year average surface water resources (SWR) amounted to $2.63 \times 10^{11} m^3$ per year, while the underground water resources (GWR) were $0.82 \times 10^{11} m^3$ per year. Consequently, the total water resource was $2.81 \times 10^{11} m^3$ per year. Additionally, the average overlapping water resources between SWR and GWR were approximately $0.64 \times 10^{11} m^3$ per year. Compared to the period 1956–1979, the national average annual water resources increased by slightly about 2% from 1980 to 2000 and declined by 3.5% from 2001 to 2017. Among the ten water resources regions, water resources in the Songhua River Basin, the Huaihe River Basin, and the Northwest Rivers Basin increased. Conversely, the remaining river basins witnessed varying degrees of decline, ranging from 1.2% to 49.6%. Under the

RegCM4 Regional Climate Model (RCM4) Medium Future Emission Scenario (RCP4.5) for greenhouse gases and aerosols, the average annual runoff in 2021–2050 may increase by 2.8% relative to 1956–2017. ③The differences in streamflow and its composition changes during the period 1956–2017 were quantified for 21 representative river basins situated across nine climatic zones and four geomorphic regions. The results indicated that while the spatial distribution of streamflow generally aligned with precipitation patterns, their temporal trends diverged. Notably, in cold regions influenced by frozen soil, streamflow exhibited an increasing trend. Conversely, river basins located in the Warm Temperate Zone, characterized by intense human activities and a fragile ecosystem, experienced a significant decrease in natural streamflow. Regarding the components of streamflow, the presence of frozen soil and karst structures led to an increased proportion of river baseflow within the overall streamflow.

Acknowledgement

THIS BOOK IS PUBLISHED with financial support from the National Natural Science Foundation of China (Grant No. 52394233), the National Key Research and Development Plan (Grant No. 2021YFC3200203), and the Young Elite Scientists Sponsorship Program by CAST (Grant No. 2023QNRC001).

CHAPTER 1

Model Architecture Design and Implementation Program of the CWAM

1.1 IDENTIFICATION OF THE COMPLEXITY OF WATER YIELD MECHANISMS IN CHINA

1.1.1 Regional Heterogeneity in China

China is situated in the southeastern region of the Eurasian continent, bordered by the Pacific Ocean to the east. As of the end of 2023, China's total population was approximately 1.41 billion, with its GDP reaching 126,058 billion Yuan. The country covers 22,909 rivers with individual basin areas of over 100 km^2 and 2865 lakes with water surface areas of larger than 1 km^2, as illustrated in Figure 1.1. Geographically, China exhibits a step-like topography that descends from the high western regions to the lower eastern plains. China's topography is high in the west and low in the east, with a step-shaped distribution. The western part of the country is predominantly characterized by mountains and plateaus, while the eastern region is primarily composed of plains.

To facilitate comprehensive water resource management, China is further divided into 10 Class I water resources regions (WRRs), 80 Class II WRRs, and 210 Class III WRRs (Figure 1.2). The ten Class I WRRs include Songhuajiang River Basin (SRB), Liaohe River Basin (LRB), Haihe River Basin (HRB), Yellow River Basin (YRB), Huaihe River Basin (HURB), Yangtze River Basin (YZRB), Southeast River Basin (SERB), Pearl River Basin (PRB), Southwest River Basin (SWRB), and Northwest River Basin (NWRB).

The country encompasses an extensive territory that spans multiple climatic-hydrological zones and diverse geological and geomorphological units. This diversity leads to considerable spatial variability in water and thermal conditions, as well as geological features,

DOI: 10.1201/9781003646648-1

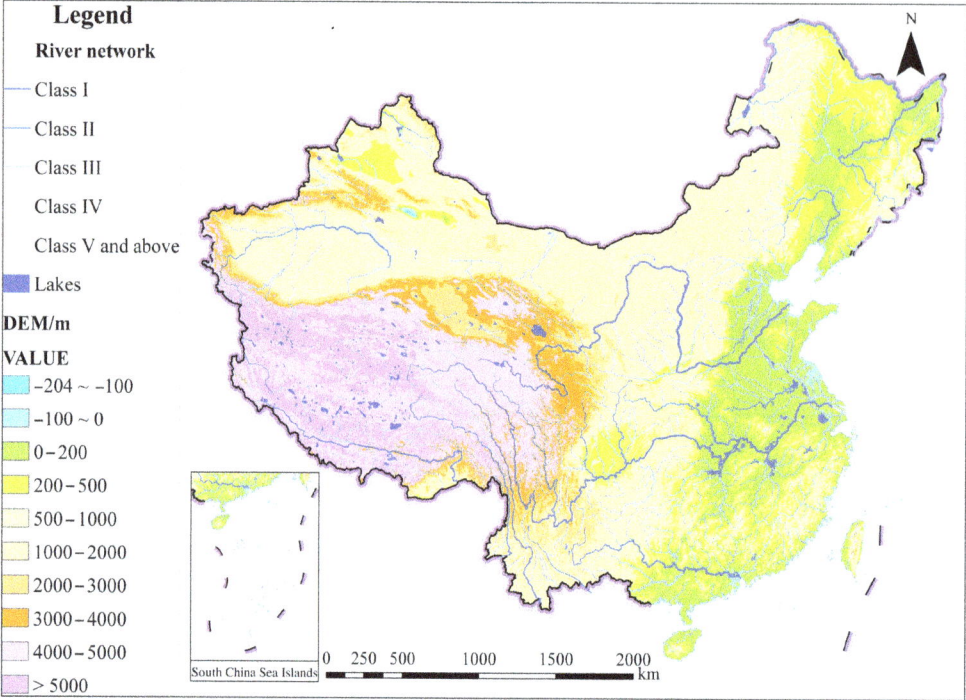

FIGURE 1.1   Topography and main rivers of China.

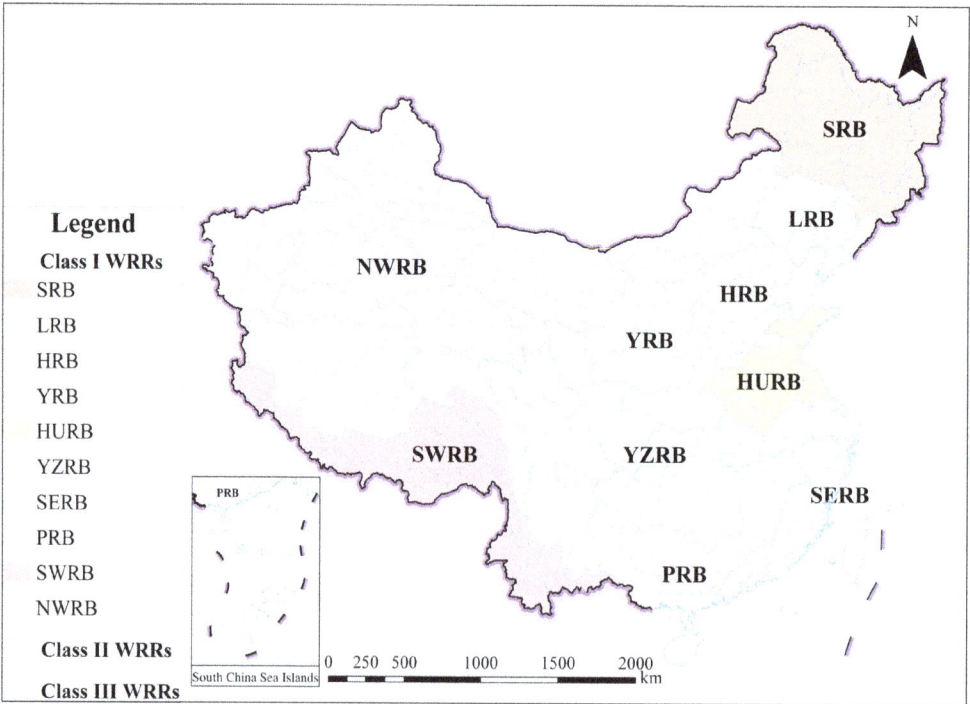

FIGURE 1.2   Water resources regions of China.

which traditional hydrological models for river basins fail to fully capture. Unlike a single river basin, China, as a large-scale region, exhibits the following characteristics.

1. It covers as many as ten climatic zones: Frigid Temperate Zone (FTZ), Median Temperate Zone (MTZ), Warm Temperate Zone (WTZ), Alpine Frigid Zone (AFZ), Alpine Subfrigid Zone (ASZ), Plateau Temperate Zone (PTZ), North Asian Tropical Zone (NATZ), Middle Asian Tropical Zone (MATZ), South Asian Tropical Zone (SATZ), and Marginal Tropical Zone (MTPZ). The water and thermal conditions vary greatly in different climatic zones. In the southern regions characterized by a humid climate, the saturation-excess runoff occurs regularly during rainfall events. Conversely, in the northern regions, the climate is dry, and the unsaturated zone of soil is thick, so infiltration-excess runoff is more common. In the case of semi-arid and semi-humid regions of the Huang-Huai-Hai River basin, both saturation-excess and infiltration-excess runoff exist. As a result, it is difficult to characterize the composite processes of saturation-excess and infiltration-excess runoff across the various climatic zones.

2. It has tens of thousands of rivers, each varying in shape, basin area, and network density. Moreover, China's northwest inland area occupies nearly one-third of the country, with many rivers flowing into inland lakes or disappearing into deserts. The surface flow model proposed by Jenson and Domingue (1988) is conventionally used to obtain the river networks and basins by setting a single sub-basin area threshold. Dealing with the complex river network structure of China, the method proposed by Jenson and Domingue (1988) exhibits certain limitations in accurately describing the boundaries of large-scale regions, determining outlets of the river basins, and identifying the rivers in inland regions.

3. It is a mountainous country, with approximately two-thirds of its landmass covered by mountains. Influenced by high topography change, the water and thermal conditions as well as vegetation show significant vertical zonal heterogeneity. Sevruk (1997) found that there exists a strong correlation between precipitation and elevation in large-scale mountainous regions. Furthermore, as elevation increases, vegetation type gradually changes from deciduous broadleaf forests to coniferous forests, then to alpine meadows, and ultimately to areas covered by glaciers and snow. This heterogeneity influences multiple model parameters, including leaf area index (LAI), fractional vegetation cover (FVC), canopy interception, and depression storage depth. Consequently, it is imperative to conduct a more precise examination of the vertical variations in model inputs and parameters.

4. It has a complex and diverse vadose zone structure that determines the redistribution of precipitation. Many rivers in China originate from highlands and mountainous regions where the soil is exposed to freezing temperatures (Li et al., 2008). The phase transition of water during freezing and thawing cycles significantly alters soil water potential, which affects the evapotranspiration and infiltration rates (Watanabe and

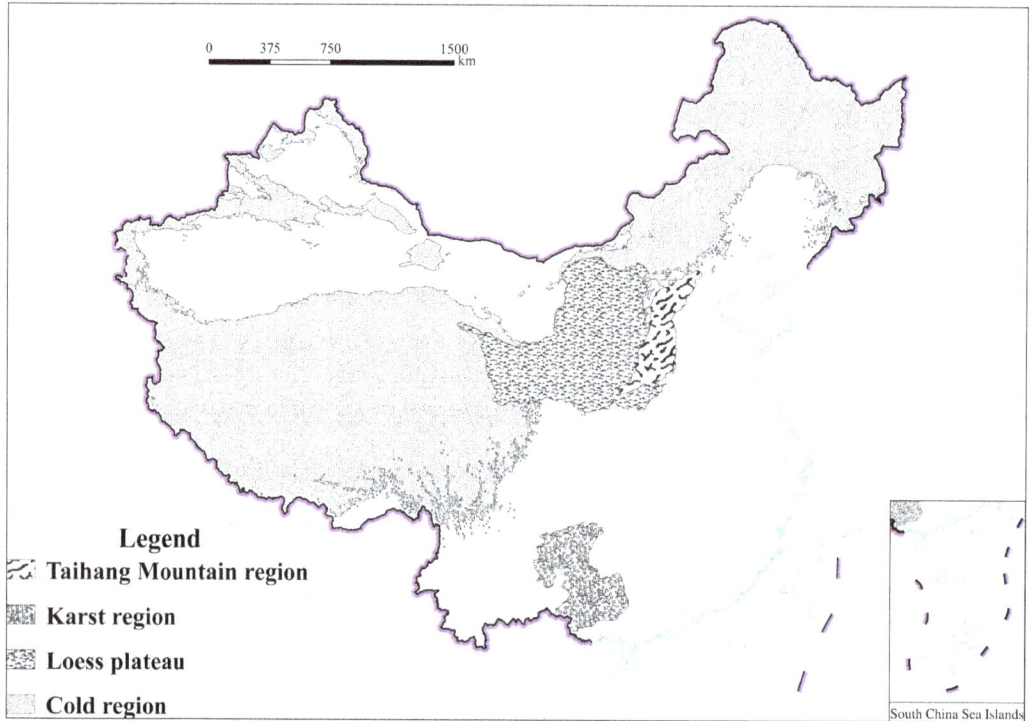

FIGURE 1.3   Schematic diagram of the distribution of the special vadose zone structures in China.

Osada, 2017). Simultaneously, one of the largest, continuous karst regions in the world is located in the Yunnan-Guizhou Plateau of southwest China (Zhang et al., 2011). Karst system has strong heterogeneity due to the existence of micropores, small fissures, and large fractures and conduits (Perrin et al., 2003; Hartmann et al., 2014). Moreover, swelling soils, as special soil types, are widely distributed all around the Loess Plateau in China. Swelling deformation occurs when swelling soils absorb water, affecting the distribution of soil pores and moisture movement (Su, 2010). Accurate description of the spatial heterogeneity of vadose zone structures is one of the important challenges to be solved. Figure 1.3 shows the spatial distribution of these special vadose zone structures in China. The karst region was determined with reference to the South China karst topographic map in National Natural Atlas of China. The Loess Plateau was from the Geographic Data Sharing Infrastructure at the College of Urban and Environmental Science, Peking University (http://geodata. pku.edu.cn). The cold region referred to Chen et al. (2005). It should be noted that Taihang Mountain Region is a typical soil-bedrock structure, which has been characterized in the WEP model.

1.1.2 Factors Affecting Water Yield in Large-Scale Regions

The water yield in large-scale regions under natural conditions is a complex and non-linear process, which is significantly influenced by climatic characteristics, subsurface conditions,

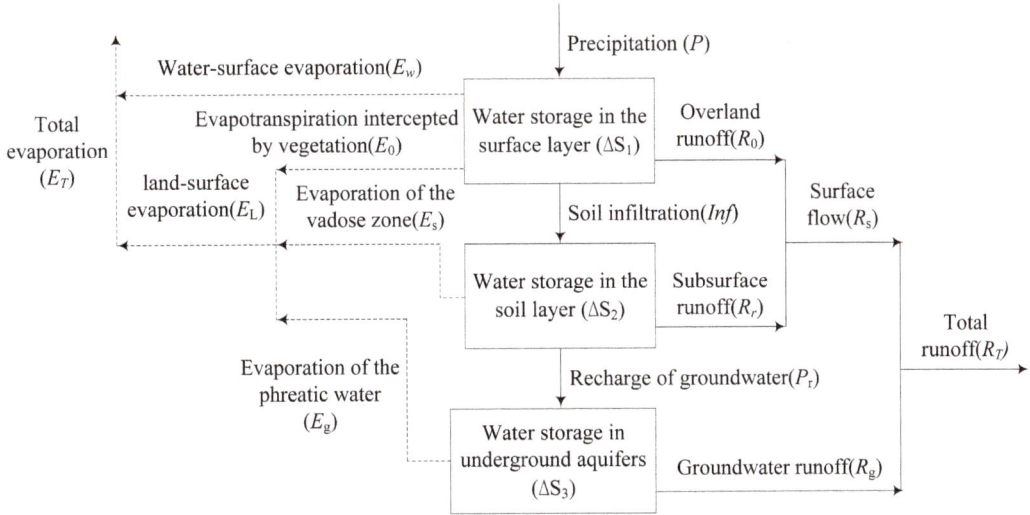

FIGURE 1.4   Schematic diagram of water yield processes.

and human activities. A schematic diagram illustrating the process of water yield in river basins is shown in Figure 1.4.

1.1.2.1 Climatic Conditions

Climate is the most basic and important factor affecting water yield in a river basin. It encompasses elements such as precipitation, air temperature, wind speed, relative humidity, and radiation. It is generally accepted that water yield is primarily governed by infiltration-excess runoff in river basins with a dry climate, while saturation-excess runoff predominates in river basins with a wet, perennial climate. Precipitation is the basic source of water resources in river basins and a key factor reflecting the wet and dry characteristics of the climate. The characteristics of precipitation directly affect infiltration and flow yield processes; meanwhile, air temperature, wind speed, relative humidity, and radiation affect soil moisture content and alter soil infiltration by controlling evapotranspiration within the river basin. For example, according to Wang et al. (2007), air temperature and evapotranspiration (ET) capacity in the YRB showed a good positive correlation, with an increase of 1°C in air temperature increasing the ET capacity by about 5.0%–7.0%. In addition, the structure of vadose zone becomes complicated with the change of air temperature. In the alpine zone, freezing of the soil occurs when the temperature drops, forming a permafrost layer, and thawing of the permafrost when the temperature rises. The alternation of soil freezing and thawing causes constant changes in effective soil water content, porosity, saturated hydraulic conductivity, etc., forming the special soil moisture dynamics in cold regions.

1.1.2.2 Subsurface Conditions

Among the various types of surface cover that influence the process of water yield in a river basin, geology, topography, vegetation type, and growth conditions play significant roles. Geological conditions, including lithology, tectonics, the degree of joint development and weathering, and soil texture, influence precipitation infiltration, evapotranspiration, water storage, and the exchange between surface and groundwater runoff. Hard and dense rocks are unfavorable for precipitation infiltration, resulting in high runoff coefficients. Conversely, if the rocks are highly weathered or soluble, they enhance the hydraulic conductivity and water-holding capacity of vadose zone, resulting in abundant underground runoff, such as in karst areas. Surface soils can be broadly classified by texture into three types: sandy, clay, and loam. Of these, sandy soils have strong hydraulic conductivity and weak water-holding capacity, clay soils have poor hydraulic conductivity and strong water-holding capacity, and loamy soils have hydraulic conductivity and water-holding capacity somewhere in between. Topography, on the one hand, affects infiltration by influencing the intensity of rainfall water supply to the subsurface. When the intensity of rainfall is the same, the soil on flat surfaces receives more water than on sloping surfaces, and thus rainfall infiltration is more effective on flat surfaces. It was found that runoff coefficients were greater in steep topographic basins than in gently sloping basins for roughly the same climatic characteristics and air inclusion zone conditions (Rui, 1996). On the other hand, the steepness of the terrain due to gravity also affects the flow rate of runoff over the subsurface, altering the timing of production and the formation of runoff volumes. According to the simulation experiment of indoor artificial rainfall, the slope gradient is in the range of 20°; when the length of the flume is certain, the rate of water flow increases gradually with the increase in the slope. There was also a significant relationship between slope gradient and the amount of water yield (Zhang et al., 2008). Vegetation intercepts and redistributes precipitation through the canopy and litter layer, increasing infiltration and evapotranspiration, slowing runoff rates, and reducing runoff volume. Studies have shown that different types of vegetation differ in their ability to intercept and redistribute precipitation, with woodlands generally being larger than shrubs and shrubs larger than grasslands (Siriwardena et al., 2006). In addition, the better the vegetation growth condition (leaf area, vegetation cover, root depth, etc.), the greater the retention and storage of precipitation in general.

1.1.2.3 Human Activities

Human activities have significantly altered local climatic and subsurface conditions, affecting the process of water yield and the water-heat balance in both direct and indirect ways. Generally, the construction of hydraulic projects, such as reservoirs, dams, and transfer projects affect the natural transport of water and directly alter the quantity and spatial and temporal distribution characteristics of natural river runoff. The construction of residential and industrial areas has resulted in hardened surfaces and a substantial increase in the percentage of impervious areas (IAs) in river basins. This change hinders the infiltration of precipitation while reducing the surface's ability to retain and store water. This resulted in a significant increase in flood flow, a shortening of the peak occurrence time, and a significant increase in the runoff coefficient. Agricultural development and soil and water conservation change soil geological conditions and local topography, thereby affecting the time-range distribution processes of precipitation and evapotranspiration.

Furthermore, the over-exploitation of water resources results in a continuous decline in groundwater levels and an increase in the thickness of vadose zones, leading to a decrease in the surface runoff coefficient and altering the interrelationship between surface water and groundwater.

1.1.3 Spatial Heterogeneity Characteristics of Water Yield Mechanisms

China has a vast area, diverse climate types, and complex subsurface conditions, which have led to significant regional differences in water production mechanisms. These variations can be summarized in the following four aspects:

1.1.3.1 Coexistence of Saturation-Excess and Infiltration-Excess Runoff

In humid climatic areas, the vadose zone can easily become saturated following rainfall, resulting in saturation-excess runoff. Conversely, in areas with a dry climate and a high thickness of the vadose zone, vegetation conditions are poor, and the precipitation is low, primarily occurring as heavy rainfall. For these areas, it is difficult for the vadose zone to become saturated by small amounts of precipitation. Therefore, the primary cause of runoff is that the intensity of rainfall exceeds the infiltration capacity of the surface soil. Of course, some arid areas may be distributed with meadows, swamps, or be recharged by snowmelt and ice melt, resulting in high soil moisture in these areas, which is easily saturated. In addition, in semi-arid and semi-humid areas, saturation-excess and infiltration-excess runoff can occur simultaneously. With changes in precipitation characteristics, soil moisture content, etc., the form of runoff yield in these areas undergoes a rapid transformation, and the precipitation-runoff relationship becomes more complex.

1.1.3.2 Coexistence of Water Yield in Cold and Non-Cold Regions

Glaciers, snow, and permafrost cause water yield mechanisms in cold regions that differ significantly from those in non-cold regions. As temperatures rise in the spring, glaciers, snow, and permafrost in cold regions thaw to recharge river runoff, which is the main reason why spring flooding occurs in many rivers. In addition, the freezing and thawing of soils in vadose zones with temperature changes creates special soil moisture dynamics in cold regions, which profoundly affects and alters the mechanism of water yield in cold regions. When the soil freezes as the temperature drops, the soil pores are closed, making it difficult for water to infiltrate. As temperatures rise and permafrost thaws, the soil's capacity to infiltrate and store water increases, and the interaction between surface water and groundwater becomes active. The rainfall-runoff coefficient for permafrost can reach up to 0.7, which is significantly higher than that of non-permafrost areas.

1.1.3.3 Coexistence of Water Yield in Mountainous Areas and Plains

The process of water yield in mountainous areas is affected by changes in elevation, in contrast to the horizontal variability observed in plains areas. In alpine areas where the terrain is highly undulating, water and heat conditions, vegetation types, and growth are characterized by vertical variations. Sevruk et al. (1997) pointed out that in large and mesoscale mountains, precipitation is influenced by topographic elements such as elevation, slope of the subsurface, and slope direction. Among these factors, the relationship between precipitation and elevation is the closest. As altitude increases, the vegetation type transitions from deciduous broadleaf forests to coniferous forests, then to alpine meadows, and finally to glacial snow cover, resulting in vertical changes in vegetation parameters such as LAI and vegetation cover. Consequently, there are large differences in the hydrothermal state and vegetation characteristics of different elevation zones, which directly affect the process of water yield such as evapotranspiration and infiltration.

1.1.3.4 Coexistence of Water Yield in Soil and Rock Structure

There are significant differences in the lithology, thickness, and structure of vadose zones that affect the process of water yield. For mountainous areas composed of soluble rocks, such as carbonate rocks, the karst development within vadose zones is strong, resulting in the formation of numerous caverns, dissolution fissures, and lysimetric holes, which makes the recharge and discharge of groundwater very rapid. For non-karst mountainous areas, the bedrock of vadose zones generally has rock-forming fissures, tectonic fissures, and weathering fissures. Influenced by the depth, size, number, and connectivity of rift development, the structure of vadose zones has strong spatial heterogeneity. In addition, in loess areas, there exists a special structure such as large pores and vertical joints, where the soil swells with water absorption and shrinks with water loss, thereby affecting the distribution of soil pores and the process of water movement. The vadose zones and aquifers in the plains consist of huge thicknesses of loose Quaternary sediments with good water storage conditions. Groundwater recharge is mainly based on precipitation infiltration.

1.2 FUNCTION POSITIONING AND SELECTION OF THE CHINA WATER ASSESSMENT MODEL (CWAM)

1.2.1 Model Functional Requirements

The large-scale region has a vast area and distinctive features and is subject to the combined influence of various climates, topographies, vegetation, and the structure of vadose zones. The mechanisms governing water yield exhibit both diversity and complexity. However, traditional distributed hydrological models for river basins are mostly for complete basins and are limited to the extent of the study area and the types of climate and subsurface conditions they can cover. Consequently, the traditional hydrological model is not sufficient for national-scale water resources assessment in four main aspects of model construction: calculation unit division, model input, spatial parameterization, and simulation methods.

In order to achieve a refined simulation of hydrological processes and a dynamic assessment of water resources in the country, the model should possess the following functions:

1. It is capable of efficiently and rapidly dividing computation units and encoding their topological relationships for multiple river basins with independent catchment relationships in a large-scale region. This ensures that the water balance and the paths of converging water flows within the computation units are not distorted during large-capacity simulations.

2. It is able to realize the coupled simulation of hydrothermal processes to adapt to the complex water and energy conditions present under the diverse climatic characteristics nationwide. Water cycle and heat cycle processes interact and are closely related, especially in cold regions.

3. It is capable of simultaneously portraying both saturation-excess and infiltration-excess runoff, to reflect the complex characteristics of water yield under varying climatic and geological conditions in this country. Due to the spatial and temporal variability of rainfall and the structural diversity of vadose zones, coupled with the impact of human activities, the water yield mechanism is quite complex, involving a variety of forms, such as hyper-permeable surface runoff, saturated surface runoff, runoff in the soil, and groundwater runoff.

4. It is able to make full use of the existing multi-source data to complete the model inputs and the effective assignment of parameters on the computation units to reflect the spatial and temporal variability of meteorological, hydrological, soil, vegetation, and other information on a nationwide scale. Model inputs and parameters are critical in determining the accuracy of model simulations. However, they vary significantly with climate characteristics and physical geographic conditions, affecting the hydrological response of the river basin.

5. It is capable of simulating hydrological processes such as evapotranspiration, evaporation, snowmelt, infiltration and soil water movement, groundwater movement and its exchange with rivers, and slope and river confluence based on the principle of balance of water, energy, and momentum, in order to support the quantitative analysis of hydrological variables, water resources, and their spatial and temporal evolution.

1.2.2 Investigation and Selection of Large-Scale Hydrological Models

Hydrological models, especially distributed physical hydrological models, are seen as a product of the combination of water yield and confluence theory and computer technology. Wagener et al. (2004) argued that any hydrological model can be expressed as equation (1.1):

$$Q = M_L(\theta_L / I) + \varepsilon_L \qquad (1.1)$$

where Q is the simulated flow at the river basin outlet; M_L and θ_L are the model structure and parameters, respectively; I is the model input; and ε_L is the local error.

It can be seen that the construction of a hydrological model is to determine a reasonable set of model structure and parameters to provide a scientific description of the natural geographic conditions of a particular catchment/region and its corresponding processes of water yield and confluence. Therefore, a model with appropriate parameters and structure needs to be selected as the basis for the design and development of a distributed hydrothermal coupling model for large-scale regions. Based on the existing studies, nine (semi-) distributed hydrological models that are currently more widely used at home and abroad were investigated. Among them, the models developed by foreign scholars include TOPMODEL

(Kirkby and Beven, 1979), SHE (Abbott et al., 1986a; Abbott et al., 1986b), IHDM (Calver, 1988), TOPKAPI (Ciarapica and Todini, 2002), VIC (Liang et al., 1994), and SWAT (Arnold and Fohrer, 2005); the models developed by Chinese scholars include DTVGM (Xia et al., 2004), GBHM (Wang et al., 2006), and WEP-L (Jia et al., 2006). These models are analyzed in terms of computation unit divisions, water yield mechanisms, confluence calculations, structural characteristics, and scope of application, as listed in Table 1.1.

TABLE 1.1   Comparison of Commonly Used Semi-distributed and Distributed Hydrological Models

Hydrological Models	Computation Unit Divisions	Water Yield Mechanisms	Confluence Calculations	Structural Characteristics	Scope of Application
TOPMODEL	Grids/sub-basins	Saturation-excess runoff	Slope runoff hysteresis function and channel evolution function	Based on topography and variable flow area theory; effects of precipitation, evaporation, etc. not considered	Used in uninformative areas, suitable for river basins with high topographic relief
SHE	Grids	One-dimensional Richards equation	Continuity and momentum equations in the St. Venant equations	Outstanding physical mechanism; complex structure and cumbersome calculations	Soil slopes, but difficult to apply to large-scale river basins
IHDM	Geomorphological unit	Two-dimensional Richards equation	One-dimensional wave equation of motion	River basin is divided into a number of drop channels and representative slopes	Suitable for simulation of storm flow in small catchments
TOPKAPI	Grids	Saturation-excess runoff	Wave equation of motion	The model parameters are strongly influenced by the grid scale, and the water movement is described by three "structurally similar" non-linear equations.	The scale is in the tens to thousands of square kilometers
VIC	Grids	Saturation-excess and infiltration-excess runoff	Instantaneous unit line method, Brooks - Corey formula, Arno model	Soil water storage capacity area distribution curve; infiltration capacity area distribution curve	Large-scale regions
SWAT	Sub-basins	CSC curve method, Green-Ampt model, Manning's formula	Muskingum method	Not suitable for detailed modeling of single flood processes	Water quality and quantity modeling in river basins at different scales
DTVGM	Grids/sub-basins	Water balance	Wave equation of motion	Integrating Hydrological Systems Theory with Distributed Modeling	River basins with a lack of hydrological information
GBHM	Geomorphological unit	One-dimensional Richards equation	One-dimensional wave equation of motion	Discrete method using unit lines combined with terrain properties	Small- or medium-scale catchment
WEP-L	Contour belts	Theory of various source areas	One-dimensional wave equation of motion	Coupled simulation of water and energy exchange processes, reflecting the theory of variable water area	Large-scale river basin

Different hydrological models have unique characteristics due to the different modeling objectives, ideas, and structures. Considering the spatial variability of water yield mechanisms nationwide, we choose to design and develop the CWAM based on the WEP-L model. The reason for this is that the WEP-L model structure has advantages in water resources assessment in large-scale river basins and has been successfully applied in the YRB and other areas.

1.2.3 WEP-L Model and Dynamic Water Resources Assessment

The water and energy transfer processes (WEP) model (Jia et al., 2001) was developed by combining the merits of physically based spatially distributed (PBSD) hydrological models and soil vegetation atmosphere transfer (SVAT) models. The model has been successfully applied in several river basins in Japan, Korea, and China with different climate and geographic conditions (Jia and Tamai, 1998; Jia et al., 2001, 2002, 2005). The WEP model has the following main characteristics: (1) combined modeling of hydrological processes and energy transfer processes, (2) consideration of the land use heterogeneity inside a computation unit by adopting the mosaic method (Avissar and Pielke, 1989), and (3) incorporation of the runoff generation theory of various source areas (Hewlett, 1982) into the model through a numerical simulation in groundwater/subsurface water flow to directly reflect topography's effects in runoff generation, thus capable of modeling infiltration excess, saturation excess, and mixed runoff generation mechanism.

To make the WEP model applicable for water resources assessment in large river basins like the YRB, the following main improvements were performed and the WEP-L model was established consequently: (i) instead of grid cells, contour belts inside small sub-basins, which were obtained based on a digital elevation model (DEM) of 1-km resolution and the area of every sub-basin less than 100 km^2, were used as computation units, and the Pfafstetter coding rule (Verdin and Verdin, 1999) was adopted to code subdivided river links and sub-basins to aid hydrological modeling in the large river basin; (ii) the soil-vegetation (SV) land use group in the WEP model was further divided into three groups of SV (grassland, forest, and bare soil land), irrigated farmland (IF), and non-irrigated farmland (NF) to consider cultivation and irrigation effects on hydrological processes; (iii) a water allocation and regulation method was developed and then coupled to WEP-L to model water use processes like reservoir regulation, canal diversion and water allocation in a coupling way with natural hydrological processes; (iv) spatial and temporal interpolations of social-economy and water use data were carried out; and (v) a snow melt model based on the temperature-index approach was developed to reflect the impacts of snow storage and melting on hydrological and energy processes as well as water resources.

1.3 MODEL ARCHITECTURE DESIGN OF THE CWAM

1.3.1 Model Framework

Based on the WEP-L model, the construction of the CWAM was divided into four major parts: determination of basic computation units, model inputs, spatial parameterization, simulation of water/energy cycle processes, and outputs. Furthermore, each part of the

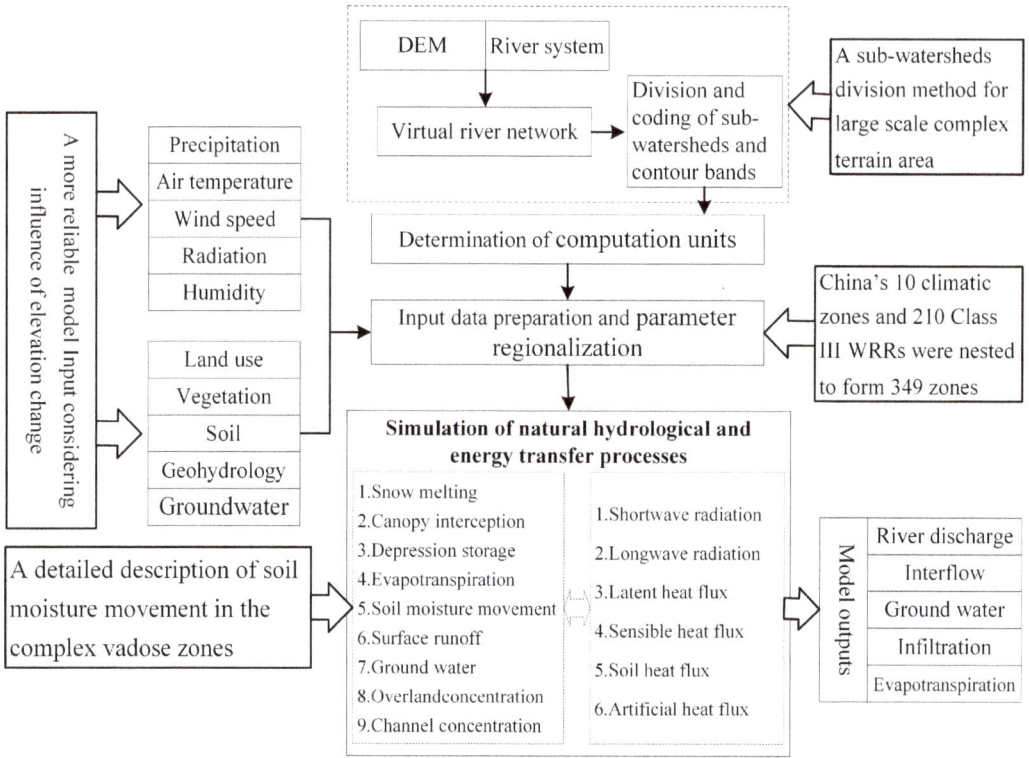

FIGURE 1.5 Schematic illustration of the architecture of the CWAM.

WEP-L model was improved to address the deficiencies in the refined simulation of large-scale hydrological processes, as shown in Figure 1.5.

1. A sub-basin division method was proposed based on automatic recognition of river basin outlets and fusion of river network with variable sub-basin area thresholds. The method mainly included the determination of river basin outlets as well as the maximum and minimum area thresholds of sub-basins, multi-threshold virtual river network integration, and sub-basin division and codification. It considered both computation efficiency and accuracy of sub-basin division, which provides an efficient basis for developing a distributed hydrological model in large-scale regions. The related study has been published in Advanced Engineering Sciences (Liu et al., 2019). Details were set out in Chapter 2.

2. Meteorological data, such as precipitation and air temperature, and vegetation data, such as LAI and coverage, are key inputs to hydrological models. Such data are significantly affected by topographic relief. Based on China's 80 Class II WRRs, the regression relationships between precipitation and air temperature with elevation were established in each WRR. Furthermore, instead of the original plane interpolation method, a three-dimensional interpolation method considering elevation effects was adopted to interpolate precipitation and air temperature. Vegetation data of different

elevation zones were obtained using the remote sensing/geographic information system (RS/GIS) techniques. Details were set out in Chapter 2.

3. The vadose zone was simply characterized as a soil-bedrock structure in the WEP model, where soil moisture movement parameters were only determined based exclusively on the texture information of different soil types (Jia et al., 2006). However, these parameters failed to reflect the soil moisture dynamics of the special vadose structure in the karst region, Loess Plateau, and cold region. Therefore, before the development of the CWAM, we carried out experimental observations and modeling of the infiltration-produced flow processes under these special conditions to better understand these phenomena. Details were set out in Chapters 3–5, respectively. To take into account the simulation efficiency of the CWAM, we focused on the soil moisture movement in the special vadose zones on the large-scale regions and simplified the simulation process of the previous small-scale regions. The effects of soil swelling deformation, karst development, and soil freezing-thawing on soil moisture movement were quantitatively analyzed, and the related parameters were modified.

4. Regarding the calibration and validation of the national model, it is very important to achieve efficient parameter regionalization. There are three categories and dozens of parameters in the WEP model, and the methods for evaluating initial values of parameters have been described in detail by Jia et al. (2006). Furthermore, considering spatial heterogeneity of meteorological hydrology and underlying surface conditions, China's 10 climatic zones and 210 Class III WRRs were nested to form 349 zones. These zones were used as the parameter tuning units in the CWAM. Details were set out in Chapter 6. Based on the verified CWAM, we have carried out a systematic analysis of water resources and their spatial variability across the country and its different climatic and geomorphic regions. Details were set out in Chapters 7 and 8.

1.3.2 Basic Structure and Modeling Approaches for Main Processes of the WEP-L Model

1.3.2.1 Model Basic Structure

The vertical structure of WEP-L within a contour belt was shown in Figure 1.6, and the horizontal structure within a sub-basin was shown in Figure 1.7. Land use was divided into five groups within a contour belt, namely, SV group, NF group, IF group, water body (WB) group, and IA group. The SV group was further classified into bare soil land, tall vegetation (forest or urban trees), and short vegetation (grassland). The IA group included impervious urban cover, urban canopy, and rocky mountains. The areal average of water and heat fluxes from all land uses in a contour belt produced the averaged fluxes in the contour belt. For pervious groups of SV, NF, and IF, nine vertical layers, namely, an interception layer, a depression layer, three upper soil layers, a transition layer, an unconfined aquifer, an aquitard, and a confined aquifer, were included in the model structure.

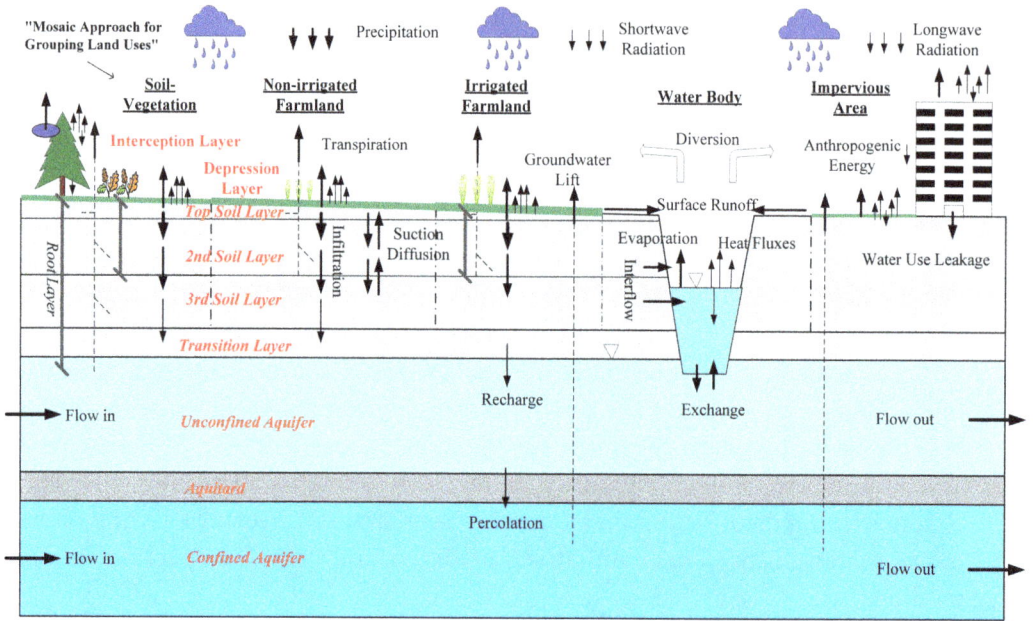

FIGURE 1.6 Vertical structure within a contour belt of the CWAM.

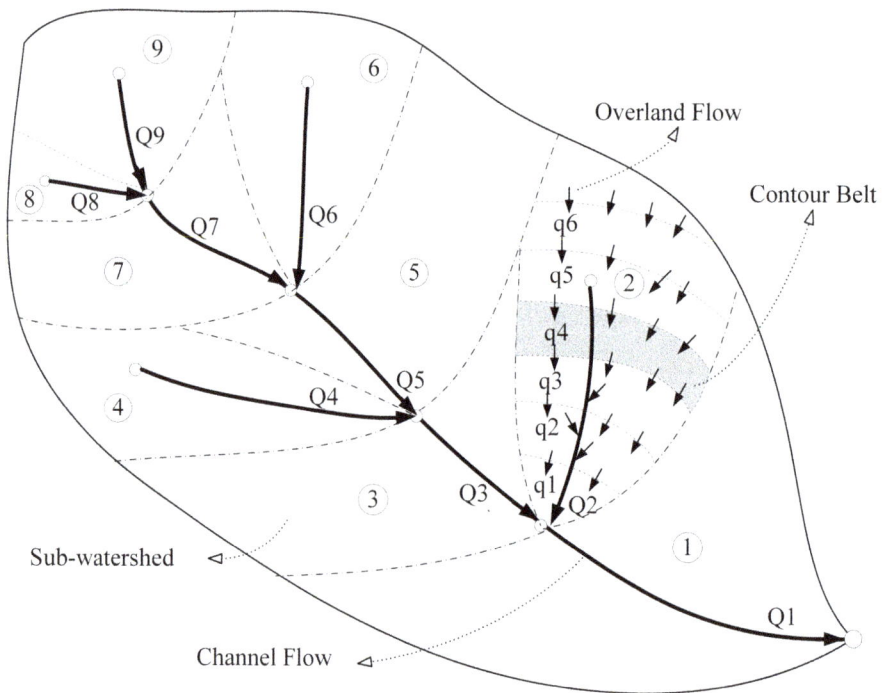

FIGURE 1.7 Horizontal structure within a sub-basin of the CWAM.

1.3.2.2 Model Calculation Methods

(1) Calculation of Energy Processes   The energy cycle directly affects and constrains water cycle processes such as evapotranspiration, snowmelt, and soil freezing and thawing in the river basin, as shown in Figure 1.8. The energy balance formulas include equations (1.2) and (1.3):

$$RN + A_e = LE + H + G + P_L + A_d \tag{1.2}$$

$$RN = RS - \alpha RS + RLD - RLU \tag{1.3}$$

where RN is the net radiation; LE is the latent heat flux; H is the sensible heat flux; G is the heat transfer in the ground; A_e is the artificial radiation; PL is the plant uptake (about 2% of the RN); A_d is the shifting term (which is often neglected except for oases and lakes); RS is the shortwave radiation reaching the surface; RLD and RLU are the atmosphere-to-surface and surface-to-atmosphere longwave radiation, respectively; α is the shortwave emissivity; the units of radiation and heat flux are MJ/m², and α is the shortwave emissivity; the units of radiation and heat flux are MJ/m².

The simulated energy transfer processes included short-wave radiation, long-wave radiation, latent heat flux, sensible heat flux, and soil heat flux.

The formulas for calculating daily short-wave radiation include equations (1.4)–(1.9):

$$RS = RS_0 \left(a_s + b_s \frac{n}{N} \right) \tag{1.4}$$

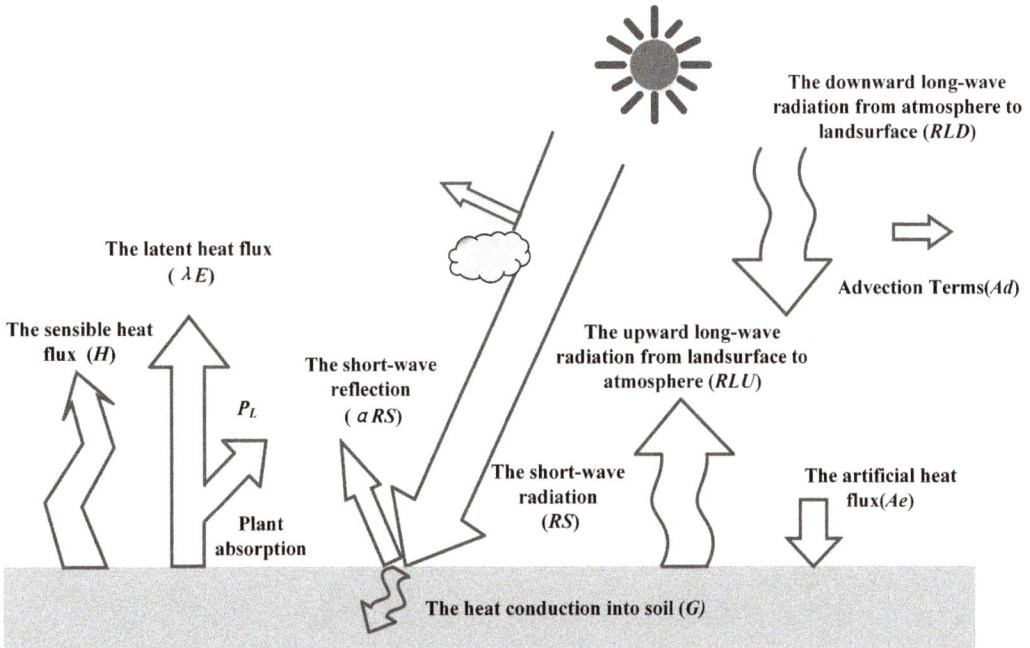

FIGURE 1.8   Schematic of land-atmosphere energy exchange.

$$RS_0 = 38.5d_r \left(\omega_s \sin\phi \, \sin\delta + \cos\phi \, \cos\delta \, \sin\omega_s \right) \tag{1.5}$$

$$d_r = 1 + 0.33\cos\left(\frac{2\pi}{365} J \right) \tag{1.6}$$

$$\omega_s = \arccos(-\tan\phi \tan\delta) \tag{1.7}$$

$$\delta = 0.4093\sin\left(\frac{2\pi}{365} J - 1.405 \right) \tag{1.8}$$

$$N = \frac{24}{\pi} \omega_s \tag{1.9}$$

where RS_0 is the short-wave radiation outside the atmosphere, and the unit of daily radiation and heat flux is MJ/m²/day if not specified; as is the short-wave radiation diffusion constant (generally 0.25); b_s is the direct short-wave radiation constant (generally 0.5); n is the number of hours of sunshine; N is the number of hours of possible sunshine; d_r is the relative distance between the sun and the earth; w_s is the solar time angle at sunset; φ is the dimension of the observation point (positive in the northern hemisphere and negative in the southern hemisphere); δ is the solar inclination; J is the number of Julian days (from 1 January of each year); and the rest of the parameters have the same significance as above.

The formulas for calculating daily long-wave radiation include equations (1.10)–(1.12):

$$RLN = RLD - RLU = -f\varepsilon\sigma(T_a + 273.2)^4 \tag{1.10}$$

$$f = a_L + b_L \frac{n}{N} \tag{1.11}$$

$$\varepsilon = -0.02 + 0.261\exp(-7.77 \times 10^{-4} T_a^2) \tag{1.12}$$

where f is the cloud impact factor; ε is the net emissivity; σ is the Stefan-Boltzmann constant (taking the value of 4.903×10^{-9} MJ m⁻² K⁻⁴ day⁻¹); T_a is the average daily air temperature (°C); a_L is the constant of the long-wave emissivity (0.1 in arid regions, and 0.3 in humid regions); b_L is the constant of the long-wave emissivity (0.9 in arid regions, and 0.7 in humid regions); the rest of the parameters have the same meaning as above.

The formulas for calculating latent heat include equations (1.13) and (1.14):

$$LE = L \times E \tag{1.13}$$

$$L = 2.501 - 0.002361 T_s \tag{1.14}$$

where L is the latent heat of water (MJ/kg); T_s is the ground surface temperature (°C); and E is evapotranspiration (mm).

The formula for calculating sensible heat is expressed as equation (1.15):

$$H = \frac{\rho_a C_P (T_s - T_a)}{r_a} \tag{1.15}$$

where ρ_a is the air density (kg/m³); C_P is the constant pressure specific heat of air (J/kg/°C); r_a is the aerodynamic impedance (s/m); the rest of the parameters have the same meaning as above.

The formula for calculating heat conduction into soil is expressed as equation (1.16):

$$G = \frac{c_s d_s (T_2 - T_1)}{\Delta t} \tag{1.16}$$

where G is the heat conduction into soil (MJ/kg); d_s is the thickness of the affected soil layer (m); T_1 and T_2 are the surface temperature (°C) at the beginning and end of the period, respectively; k_s is the soil heat transfer coefficient (Wm⁻¹ K⁻¹); and c_s is the soil heat capacity coefficient (MJ m⁻¹ K⁻¹).

Furthermore, the force-restore method (FRM) was used to solve the surface temperature of different land covers. This method is a better approximation to the classical heat diffusion equation compared with other methods. Hu and Islam (1995) suggested an optimal parameter α, which cannot only ensure minimum distortion of FRM to sinusoidal diurnal forcing but also make the distortion to higher harmonics negligible. They are followed in this research with equations (1.17)–(1.20):

$$\alpha \frac{\partial T_s}{\partial t} = \frac{2}{c_h \cdot d_0} G - w(T_s - T_d) \tag{1.17}$$

$$\frac{\partial T_d}{\partial t} = \frac{1}{\tau}(T_s - T_d) \tag{1.18}$$

$$\alpha = 1 + 0.943\left(\frac{\delta}{d_0}\right) + 0.233\left(\frac{\delta}{d_0}\right)^2 + 0.0168\left(\frac{\delta}{d_0}\right)^3 - 0.00527\left(\frac{\delta}{d_0}\right)^4 \tag{1.19}$$

$$d_0 = \sqrt{2k_h/(c_h w)} \tag{1.20}$$

where G is the heat conduction into soil; T_s is the surface temperature; T_d is the deep soil temperature (approximated as the daily average of T_s), v is the considered soil depth (selected as the thickness of top soil layer); d_0 is the damping depth of the diurnal temperature wave, k_h is the soil heat conductivity, c_h is the soil volumetric heat capacity; $\omega = 2\pi/\tau$ and $\tau = 86400$. The soil thermal properties depend on the water content and the mineral composition of the soil.

(2) Calculation of Hydrological Processes   The simulated hydrological processes included snow melting, evapotranspiration, infiltration, surface runoff, subsurface runoff, groundwater flow, overland flow, and river flow.

① *Evapotranspiration*   Evapotranspiration in a contour belt consists of interception of vegetation canopies (evaporation from the wet part of leaves), evaporation from WB, soil, urban cover, and urban canopy, and transpiration from the dry fraction of leaves, with the source from the three upper soil layers. The averaged evapotranspiration E is expressed as equation (1.21):

$$E = F_W E_w + F_{sv} E_{sv} + F_U E_U \qquad (1.21)$$

where, F_W, F_{SV}, and F_U are the area fractions of WB, SV, and IA, respectively; E_W, E_{SV}, and E_U are the evaporation or evapotranspiration from them, respectively.

The evaporation from the WB or the ponded water in the depression storage was calculated using the Penman equation (1.22):

$$E_w = \frac{(RN-G)\Delta + \rho_a C_p \delta_e / r_a}{\lambda(\Delta+\gamma)} \qquad (1.22)$$

where RN is the net radiation; G is the heat conduction into the soil; Δ is the gradient of saturated vapor pressure to temperature; δ_e is the air vapor pressure deficit, r_a is the aerodynamic resistance, ρ_a is the air density; C_p is the air specific heat; λ is the latent heat of the water and the psychometric constant.

The evaporation from the IA was taken as the smaller one of current depression storage and the potential evaporation. In this study, the maximum depression storage of IA is assumed as 5 mm.

The evapotranspiration from the SV group is calculated as equation (1.23):

$$E_{sv} = E_{i1} + E_{i2} + E_{tr1} + E_{tr2} + E_s \qquad (1.23)$$

where E_i is the interception of vegetation; E_{tr} is the transpiration from the dry part of vegetation leaves, with numbers 1 and 2 representing tall vegetation and short vegetation, respectively, and E_s is the evaporation from soils.

The computation of interception is referred to the Noilhan and Planton (1989) model, which is an interception reservoir method. The evaporation from soil is assumed to come only from the topsoil layer. The formulas include equations (1.24)–(1.28):

$$E_i = Veg \cdot \delta \cdot E_v \qquad (1.24)$$

$$\partial W_r / \partial t = Veg \cdot \delta \cdot E_v \qquad (1.25)$$

$$R_r = \begin{cases} 0 & W_r \le W_{r\max} \\ W_r - W_{r\max} & W_r > W_{r\max} \end{cases} \qquad (1.26)$$

$$\delta = (W_r - W_{r\max})^{2/3} \qquad (1.27)$$

$$W_{r\max} = 0.2 \cdot Veg \cdot LAI \qquad (1.28)$$

where Veg is the fraction of tall (or short) vegetation in the SV group; δ is the fraction coefficient of the foliage covered by a water film; E_v is the potential evaporation on vegetation surface; W_r is the storage of the interception reservoir; W_{rmax} is the maximum W_r; P is the precipitation; R_r is the drainage rate from the canopy when W_r exceeds W_{rmax} and LAI is the leaf area index.

The actual transpiration was calculated using the Penman–Monteith equation (Monteith, 1973) and the canopy resistance (Noilhan and Planton, 1989), which is related to the soil moisture condition. The formulas include equations (1.29) and (1.30):

$$E_{tr} = Veg \cdot (1-\delta) \cdot E_{PM} \tag{1.29}$$

$$E_{PM} = \frac{(RN-G)\Delta + \rho_a C_p \delta_e / r_a}{\lambda[\Delta + \gamma(1 + r_c/r_a)]} \tag{1.30}$$

where E_{PM} is the Penman–Monteith transpiration; r_c is the canopy resistance, and the others as denoted above.

The transpiration is supplied from the soil layers by the roots. A root uptake model is adopted, which assumes that the root uptake intensity linearly decreases with the increase of root depth, and the uptake in the upper half root zone accounts for 70% of the total uptake. The transpiration of tall vegetation is assumed to originate from the three upper soil layers in Figure 1.6, while that of short vegetation originates from only the two upper ones.

The aerodynamic resistance r_a under neutral atmospheric conditions can be represented as equation (1.31) (Monteith, 1973):

$$r_a = \frac{\ln\left[(z-d)/z_{om}\right] \cdot \ln\left[(z-d)/z_{ox}\right]}{\kappa^2 U} \tag{1.31}$$

where z is the measurement height of wind speed, humidity, and temperature; κ is the von Karman constant; U is the wind speed; d is the displacement height; z_{om} is the roughness height of momentum, and $z_{ox} = z_{om}$ for momentum transfer, z_{ov} (the roughness height of vapor) for vapor transfer or z_{oh} (the roughness height of heat) for heat transfer, respectively. The Monin–Obukhov similarity theory is used to modify the computation of aerodynamic resistance (Brutsaert, 1982) under unstable and stable atmospheric conditions.

The canopy resistance r_c is also called the surface resistance. Noilhan and Planton (1989) are followed to calculate it. It is a summation of the contributions of the stomatal resistance of individual leaves. The formula can be expressed as equation (1.32):

$$r_c = \frac{r_{s\min}}{LAI} f_1(T) f_2(VPD) f_3(PAR) f_4(\theta) \tag{1.32}$$

where $r_{s\min}$ is the canopy minimum stomatal resistance; LAI is the canopy leaf area index; f_1 is the dependence on the air temperature T, f_2 is the dependence on the vapor pressure deficit (VPD) of the air; f_3 is the influence of the photosynthetically active radiation flux (PAR); f_4 is the effect of the soil moisture content.

Evaporation from soils was assumed to come only from the top layer. It was usually estimated by multiplying the potential evaporation (based on the Penman equation) by an evaporation coefficient, which is called the potential method hereafter. However, the potential method may cause theoretical nonconsistency of heat flux partition on the soil surface because the net radiation and soil heat flux corresponding to the saturated vapor pressure of soil are used in the Penman equation, while the actual soil may be unsaturated. Based on the energy balance on the soil surface, aerodynamic diffusion equations of latent and sensible heat fluxes, and the wetness function concept, we derived the modified Penman equation (1.33) to compute actual soil evaporation directly.

$$E_s = \frac{(RN-G)\Delta + \rho_a C_p \delta_e / r_a}{\lambda[\Delta + \gamma(1+\gamma/\beta)]} \tag{1.33}$$

where β is the wetness function, and the other notations are the same as mentioned above.

The wetness function is defined as in equation (1.34) (Lee and Pielke, 1992) and it is estimated by equation (1.35), which is a modified version of Lee and Pielke's β-equation:

$$\beta = [e(T_s) - e(T_a)]/[e_s(T_s) - e(T_a)] \tag{1.34}$$

$$\beta = \begin{cases} 0 & \theta \leq \theta_m \\ \frac{1}{4}\left[1 - \cos(\pi(\theta - \theta_m)/(\theta_{fc} - \theta_m))\right]^2 & \theta_m < \theta \leq \theta_{fc} \\ 1 & \theta > \theta_{fc} \end{cases} \tag{1.35}$$

where $e(T_s)$ is the surface vapor pressure; $e_s(T_s)$ is the saturated surface vapor pressure, $e(T_a)$ is the air vapor pressure; T_s is the surface temperature; T_a is the air temperature; θ is the volumetric soil moisture content; f_c is the field capacity of the topsoil layer; m is the moisture content correspondent to the monomolecular suction. The difference between equation (1.34) and the Lee and Pielke's β-equation is that θ_m terms are added here. θ_m should be considered because soil evaporation cannot continue when the soil suction becomes equal to or larger than the monomolecular suction.

② *Infiltration*　Considering infiltration into a vertical uniform soil column when the surface is ponded, Green and Ampt proposed an infiltration model by assuming there is a wetting front that separates saturated soil above from soil below and by using Darcy's law. Compared with the other infiltration models, the Green-Ampt model has the advantages of simplicity, physically based characteristics, and measurable parameters. Mein and Larson (1973) extended it to model infiltration into uniform soil during a steady rain, and Moore and Eigel (1981) extended it to model infiltration into two-layered soil profiles during steady rains. Moreover, Jia and Tamai (1997) suggested a generalized Green–Ampt model for infiltration into multilayered soil profiles during unsteady rains. The generalized Green–Ampt model is summarized as follows.

Supposing that the wetting front is in the mth soil layer (see Figure 1.9), the infiltration rate can be expressed as equation (1.36):

$$f = k_m\left(1 + \frac{A_{m-1}}{B_{m-1} + F}\right) \tag{1.36}$$

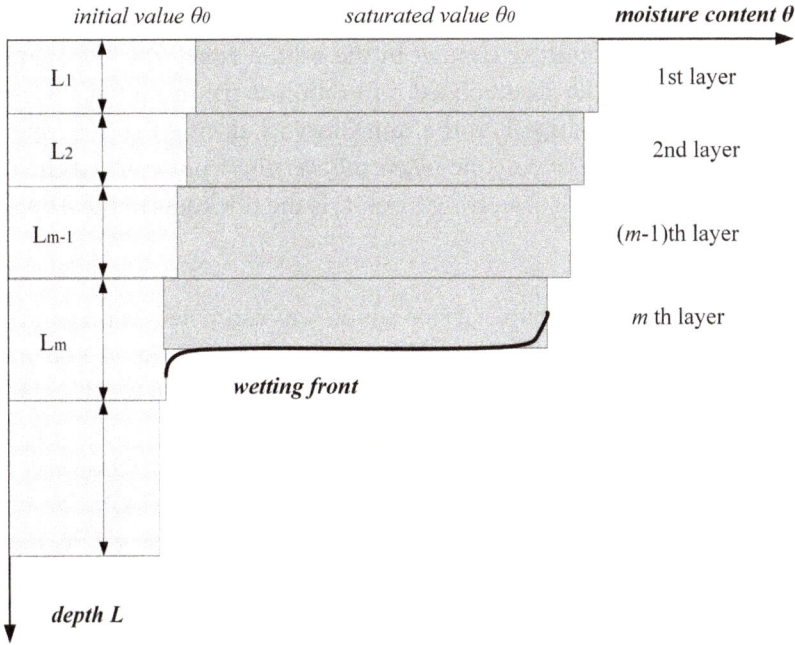

FIGURE 1.9   Schematic of infiltration into a multilayered soil profile.

where f is the infiltration rate; F is the accumulated infiltration, and the others are described later. The calculation of accumulated infiltration includes two cases. The accumulated infiltration is calculated in equation (1.37) if the surface ponding occurs with the wetting front at the $(m-1)$th layer and continues since then:

$$F - F_{m-1} = k_m(t - t_{m-1}) - A_{m-1} \cdot \ln\left(\frac{A_{m-1} + B_{m-1} + F}{A_{m-1} + B_{m-1} + F_{t_{m-1}}}\right) \qquad (1.37)$$

Whereas it is calculated using equations (1.38)–(1.41), if the surface ponding begins at the present time step t_n with no ponding at the last time step t_{n-1}:

$$F - F_p = k_m(t - t_p) + A_{m-1} \cdot \ln\left(\frac{A_{m-1} + B_{m-1} + F}{A_{m-1} + B_{m-1} + F_p}\right) \qquad (1.38)$$

$$F_p - F_{n-1} = I_p(t_p - t_{n-1}); \quad t_p = t_{n-1} + \frac{F_p - F_{n-1}}{I_p} \qquad (1.39)$$

$$A_{m-1} = \left(\sum_{1}^{m-1} L_i - \sum_{1}^{m-1} L_i k_m / k_i + SW_m\right) \cdot \Delta\theta_m \qquad (1.40)$$

$$B_{m-1} = \left(\sum_{1}^{m-1} L_i k_m / k_i\right) \cdot \Delta\theta_m - \sum_{1}^{m-1} L_i \Delta\theta_i; \quad F_{m-1} = \sum_{1}^{m-1} L_i \Delta\theta_i \qquad (1.41)$$

where SW is the capillary suction at the wetting front; k is the hydraulic conductivity in the wetted zone; θ_s is the moisture content in the wetted zone, θ_0 is the initial moisture content; t is the time; F_p is the accumulated infiltration at the instant of surface ponding; t_p is the time to surface ponding; I_p is the rain intensity during the nth time step when surface ponding occurs; t_{m-1} is the time when the wetting front reached the interface of mth and $(m-1)$th; L is the depth of wetting front; L_i is the thickness of the ith soil layer and $\Delta\theta = \theta_s - \theta_0$.

③ *Surface Runoff*   In the WB group, surface runoff was estimated as precipitation minus evaporation. In the IA group, surface runoff can be obtained by doing a balance analysis of depression storage, precipitation, and evaporation on land surfaces. It is assumed that there is no infiltration in these two groups.

In the SV group, surface runoff consists of two parts, namely, infiltration excess (Horton-type runoff) during heavy rainfall periods and saturation excess (Dunne-type runoff) during the other periods. A heavy rainfall period is defined as a period during which the rainfall intensity is larger than the saturated soil hydraulic conductivity.

The infiltration excess $R1_{ie}$ was solved by applying the generalized Green-Ampt model to infiltration in three soil layers during heavy rainfall periods. The formulas to compute infiltration excess include equations (1.42) and (1.43):

$$\frac{\partial H_s}{\partial t} = P - E_0 - f - R1_{ie} \tag{1.42}$$

$$R1_{ie} = \begin{cases} 0 & H_s \leq H_{s\max} \\ H_s - H_{s\max} & H_s > H_{s\max} \end{cases} \tag{1.43}$$

where H_s is the depression storage on the soil surface; $H_{s\max}$ is the maximum depression storage (set as 15 mm for paddy and 5 mm for other pervious land in this study); P is the rainfall; E_0 is the evaporation; f is the infiltration rate calculated with equation (1.36). In equation (1.43), the gradual variation of $R1_{ie}$ after $H_s > H_{s\max}$ is neglected because a time step of 1h is adopted in this study, which is believed to be long enough to justify the approximation. In addition, the interception storage of vegetation is assumed to be full; E_0 equals the potential evaporation, and transpiration is neglected during heavy rainfall periods.

The saturation excess $R1_{se}$ during the left periods may occur if the groundwater level in the unconfined aquifer rises and the topsoil layer becomes nearly saturated. It can be deduced by doing a balance analysis in every soil layer (the Richards model). The formulas include equations (1.44)–(1.52).

a. *Depression storage layer*

$$\frac{\partial H_s}{\partial t} = P(1 - Veg_1 - Veg_2) + Veg_1 \cdot R_{r1} + Veg_2 \cdot R_{r2} - E_0 - Q_0 - R1_{se} \tag{1.44}$$

$$R1_{se} = \begin{cases} 0 & H_s \le H_{s\max} \\ H_s - H_{s\max} & H_s > H_{s\max} \end{cases} \tag{1.45}$$

b. *Top soil layer*

$$\frac{\partial \theta_1}{\partial t} = \frac{1}{d_1}(Q_0 + QD_{12} - Q_1 - R2_1 - E_1 - Etr_{11} - Etr_{21}) \tag{1.46}$$

c. *Second soil layer*

$$\frac{\partial \theta_2}{\partial t} = \frac{1}{d_2}(Q_1 + QD_{23} - QD_{12} - Q_2 - R2_2 - Etr_{12} - Etr_{22}) \tag{1.47}$$

d. *Third soil layer*

$$\frac{\partial \theta_3}{\partial t} = \frac{1}{d_3}(Q_2 - QD_{23} - Q_3 - Etr_{13}) \tag{1.48}$$

$$Q_j = k_j(\theta_j)(j=1,2,3) \tag{1.49}$$

$$E_0 = \min\{E_p,(H_s + P_{in})E_p/(E_p + Q_p)\}; E_1 = E_s - E_0; \\ P_{in} = P(1 - Veg_1 - Veg_2) + (Veg_1 R_{r1} + Veg_2 R_{r2}) \tag{1.50}$$

$$Q_0 = \min\{Q_p,(H_s + P_{in})Q_p/(E_p + Q_p)\}; Q_p = \min\{k_1(\theta_1), Q_{0\max}\}; \\ Q_{0\max} = W_{1\max} - W_{10} - Q_1 \tag{1.51}$$

$$QD_{j,j+1} = k_{j,j+1} \cdot \frac{\psi_j(\theta_j) - \psi_{j+1}(\theta_{j+1})}{(d_j + d_{j+1})/2}; \\ k_{j,j+1} = \frac{d_j \cdot k_j(\theta_j) + d_{j+1} \cdot k_{j+1}(\theta_{j+1})}{d_j + d_{j+1}}(j=1,2) \tag{1.52}$$

In the above equations, H_s is the depression storage on soil surface; $H_{s\max}$ is the maximum H_s; Veg_1 and Veg_2 are the fraction of tall and short vegetation, respectively; R_{r1} and R_{r2} are the drainage rate from tall and short vegetation, respectively; Q is the gravity drainage; $QD_{j,j+1}$ is the suction diffusion from the $(j+1)$th soil layer to the jth layer; E_0 and E_1 are the evaporation from the depression storage layer and top soil layer, respectively; Etr is the transpiration with the first subscript representing vegetation type (1 = tall vegetation and 2 = short vegetation) and the second one representing soil layer; R_2 is the subsurface runoff; θ is the moisture content; θ_s is the saturated moisture content; $k(\theta)$ is the hydraulic conductivity correspondent to θ; $\Psi(\theta)$ is the soil suction correspondent to θ; d is the thickness of the soil layer; $W = \theta \cdot d$ is the water storage of the soil layer, W_{10} is the initial water storage of the top soil layer and the other notations are the same as mentioned above. Except where especially mentioned, the numbers or subscripts of all variables mean layer numbers with 0, 1, 2, and 3 representing the depression storage layer, top soil layer, second soil layer, and third soil layer, respectively.

The continuity of water movement was considered when the application of the generalized Green-Ampt model was switched into that of the Richards model and *vice versa*. When the application of the generalized Green-Ampt model is switched to that of the Richards model, the initial moisture contents of three unsaturated soil layers for the Richards model are computed based on the depth of the wetting front from the generalized Green-Ampt model. However, when the application of the Richards model is switched to that of the generalized Green-Ampt model, the moisture contents of the three soil layers from the Richards model provide initial values for the generalized Green-Ampt model, and no special treatment is required. It should be mentioned that there are some approximations in the application of the generalized Green-Ampt model. Though heat flux partition and surface temperature solution are also carried out, soil evaporation, canopy transpiration, and the redistribution of soil moisture below the wetting front are neglected. It is believed that these factors can be neglected during heavy rainfalls.

④ *Subsurface Runoff*   The subsurface runoff was calculated using equation (1.53), according to the land slope and the soil hydraulic conductivity.

$$k_2 = k(\theta) \cdot \sin(slope) \cdot L \cdot d \tag{1.53}$$

where R_2 is the subsurface runoff from the unsaturated soil layers; L is the channel length in the contour belt, and others as denoted above.

Snow storage and melting processes were simulated using the temperature-index approach (Maidment, 1992) on a daily basis.

⑤ *Groundwater Flow and Groundwater Outflow*   Taking account of the recharge from unsaturated soil layers and lifted groundwater as source terms, a quasi-3D simulation was performed for groundwater flow to consider the interactions between surface water and groundwater by using the Boussinesq equations (1.54)–(1.55).
unconfined aquifer:

$$C_u \frac{\partial h_u}{\partial t} = \frac{\partial}{\partial t}\left(k_u h_u \frac{\partial h_u}{\partial x}\right) + \frac{\partial}{\partial y}\left(k_u h_u \frac{\partial h_u}{\partial y}\right) + (Q_3 + WUL - RG - Per - E) \tag{1.54}$$

confined aquifers:

$$C \frac{\partial h}{\partial t} = \frac{\partial}{\partial t}\left(kD \frac{\partial h}{\partial x}\right) + \frac{\partial}{\partial y}\left(kD \frac{\partial h}{\partial y}\right) + (Per - GWP - Perc) \tag{1.55}$$

where C_u is the specific yield; C is the storage coefficient; h_u and h are the groundwater heads in the unconfined aquifer and confined aquifers, respectively; k_u and k are the hydraulic conductivities of the unconfined aquifer and confined aquifers, respectively; D is the thickness of confined aquifers; Q_3 is the recharge from unsaturated soil layers; RG is the groundwater outflow to rivers; WUL is the water use leakage; GWP is the pumped groundwater; Per and $Perc$ are the percolation to the aquifer below; E is the evapotranspiration from groundwater when the unconfined groundwater level rises above the third soil layer. $E = F_{SV} Etr_{13}$ if h_u rises to the third soil layer; $E = F_{SV} (Etr_{12} + Etr_{13} + Etr_{22})$ if h_u rises to the second soil layer or $E = F_{SV} (E_s + Etr_{11} + Etr_{12} + Etr_{13} + Etr_{22})$ if h_u rises to the topsoil layer with the notations as described above.

Groundwater outflow was calculated by equation (1.56), according to the hydraulic conductivity k_b of riverbed material and the difference between river water stage H_r and groundwater level h_u.

$$RG = \begin{cases} k_b A_b (h_u - H_r)/d_b & h_u \geq H_r \\ -k_b A_b \left[1+(H_r - Z_b)/d_b\right] & h_u < H_r \end{cases} \tag{1.56}$$

where A_b is the seepage area of the riverbed; Z_b is the elevation of the riverbed; d_b is the thickness of the riverbed material.

River flow. The river flow was routed for every sub-basin and a main channel by using the kinematic wave method, which can be expressed as equations (1.57) and (1.58).

$$\frac{\partial A}{\partial t} + \frac{\partial Q}{\partial x} = q_L \tag{1.57}$$

$$Q = \frac{A}{n} R^{2/3} S_0^{1/2} \tag{1.58}$$

where A is the area of lateral section; Q is the discharge; q_L is the lateral inflow of unit channel length; n is the Manning coefficient; R is the hydraulic radius; S_0 is the longitudinal slope of the river bed. In addition, $q_L = \Sigma[(R1 + R2 + RG + R_{sew} + Dx \cdot Dy]/L$, where Σ denotes summation of all contour belts in the sub-basin; R_{sew} is the sewerage in a contour belt of the sub-basin; Dx and Dy are the unit size in x and y directions; L is the channel length and the others as mentioned before.

⑥ *Anthropogenic Components*   The water use in contour belt was deduced by using population and water use per capita. The water use per capita was decided according to statistics of water use in a river basin. In addition, water use leakage was deduced from water use and the leakage rate of the water supply system.

The sewerage was equal to water use subtracted by leakage. It was set as one part of the lateral inflow to the channel.

The groundwater lift has twofold utilization, the drinking water and the irrigation water. The drinking water was calculated according to the annual drinking water lift and the population distribution. The irrigation water was calculated based on the annual lift, paddy area, and irrigation period.

1.3.3 Concepts and Approaches for Dynamic Assessment of Water Resources

The scope of water resources assessed in the traditional approach includes surface water and groundwater, both of which exist in gravity-driven form. However, unsaturated soil moisture in vegetation root zones and intercepted precipitation on vegetation are effective for ecology, and evaporation from the depression layer of residential areas is also effective for people-living environment because it can wet the dry air and lower the air temperature in hot summer. Thus, these parts of evapotranspiration, i.e., the precipitation directly utilized by the ecosystem, should also be considered for water resources assessment. The traditional water resources can be called "special water resources", and those including the precipitation directly utilized by the ecosystem can be called "general water resources".

The general water resources comprise the sum of specialized water resources and the precipitation directly utilized by the ecosystem, which can be calculated as equation (1.59).

$$W = (R_s + R_g) + (E_i + E_d + E_t) \tag{1.59}$$

where the variables are defined as follows: W, general water resources; R_s, surface water resources; R_g, groundwater resources non-overlapped with surface water resources; E_i, interception of vegetation canopies; E_d, evaporation from depression layers of residential and vegetation areas; and E_t, vegetation transpiration, i.e., utilization of soil moisture non-overlapped with surface water and groundwater in vegetated areas. In addition, $R_s + R_g$ is the special water resources, and $E_i + E_d + E_t$ is the precipitation utilized by the ecosystem.

1.4 IMPLEMENTATION PROGRAMME OF THE CWAM

Based on determining the basic computation units and their topological relationships, there were 38 main computation modules in the model (Table 1.2). Modules 1–31 were the inherent computation modules of the WEP-L model. To realize the demand for refined simulation of different climatic and hydrological zones and geological and geomorphological units in large-scale regions, based on the design of the model architecture, modules 32–38 were developed to address the key issues such as spatial three-dimensional spreading of meteorological data, spatial differentiation of vegetation parameter assignments, and modification of the structure of the special air-sealed zones and their moisture movement parameters, etc. The overall computation process is shown in Figure 1.10.

1.4.1 Spatial and Temporal Spreading of Daily Meteorological Data over the Sub-Basins

First, according to the actual needs or existing results, the large-scale area will be divided into a number of subzones, such as administrative zones, climatic zones, water resource zones, etc., and as far as possible, the areas with drastic changes in topography will be separated from the flat areas. Then, based on the observed data of meteorological stations and the elevation where they are located in each subzone, a univariate regression relationship of meteorological data changes with elevation can be established zone by zone to obtain the coefficient of determination and the linear rate of change. If the coefficient of determination of the subzones is greater than 0.5, it is considered that the meteorological data in

TABLE 1.2   Main Module of the CWAM

Id	Name	Function
1	WEP_Main	Control structure framework of the main program.
2	Read_IO_File	Read input and output file names and convert them to absolute paths for file invocation.
3	Open_IO_File	Open the input and output files and prepare to read and write data.
4	Define_Para	Define parameters, variables, arrays, etc. (including time parameters).
5	Read_Para	Read data and parameters.
6	Set_Init_Status	Set initial and boundary conditions (flow, water table, etc.) for calculation units.
7	Landuse_Year	Land use was calculated year by year and reclassified into five categories.
8	River_alfa	Calculation of river-related roughness coefficients, such as Manning coefficient, and overbank effect parameters.
9	Sub_radi	Calculation of daily radiation in sub-basins.
10	P_Hour	Calculation of hourly rainfall based on daily rainfall downscaling.
11	Runoff_1_water	Calculation of the water and heat fluxes of water body in contour belts.
12	Runoff_2_soilveg	Calculation of the water and heat fluxes of soil and vegetation in contour belts.
13	Runoff_3_urban	Calculation of the water and heat fluxes of impervious area in contour belts.
14	Runoff_4_svir	Calculation of the water and heat fluxes of irrigated farmland in contour belts.
15	Runoff_5_svni	Calculation of the water and heat fluxes of non-irrigated farmland in contour belts.
16	Runoff_2_gampt	Calculation of the water and heat fluxes of soil and vegetation during heavy rainfall in contour belts using the Green-Ampt model.
17	Runoff_2_svei	Calculation of the water and heat fluxes of soil and vegetation during non-heavy rainfall in contour belts using the Richards model.
18	Accu_F	Calculation of the cumulative infiltration of soil and vegetation during heavy rainfall in contour belts using the Green–Ampt model.
19	Resis	Calculation of the aerodynamic resistance and canopy resistance.
20	Potential_ET	Calculation of the potential evaporation using the Penman equations.
21	Soil_E	Calculation of the evaporation from soils using the Penman equations.
22	Root_E	Calculation of the transpiration from soil layers by roots based on a root uptake model.
23	RFRM	Calculation of the surface temperature and the heat conduction into soil using the Force-Restore methods.
24	Unsaturated_K	Calculation of the unsaturated water conductivity corresponding to different soil water content using the Mualem equation.
25	Snow_ice	Calculation of the accumulation and melting of snow and ice.
26	Gwater	Calculation of the daily groundwater movement using the finite difference method.
27	Gw_River_exchange	Calculation of the exchange between groundwater and river runoff.
28	Conflux_Overland	Calculation of the confluence of overland flow using the one-dimensional motion wave method
29	Conflux_River	Calculation of the confluence of stream flow using the one-dimensional motion wave method.
30	Output_ViewRange	Output of the daily, monthly, and annual water cycle elements (water fluxes, resource quantities) in the concerned interval.
31	Output_Basin	Output of the daily, monthly, and annual water cycle elements (water fluxes, resource quantities) in the entire basin.
32	Sub_Mete_Zone	Distribution of daily meteorological data in the sub-basins.
33	Mon_Lai_Veg	Distribution of monthly vegetation data in the sub-basins.
34	Sub_Zone	Identification of the vadose zone structure type of the sub-basin.
35	Accu_Soil_Loess	Calculation of soil water movement in the Loess Plateau region
36	Accu_Soil_Karst	Calculation of soil water movement in the karst region
37	Accu_Soil_Frozen	Calculation of soil water movement in the cold region
38	Para_Soil_Air	Calculation of precipitation infiltration under the influence of air resistance.
39	ZoneCoef_Para	Partition setting of parameters to be debugged of the model.

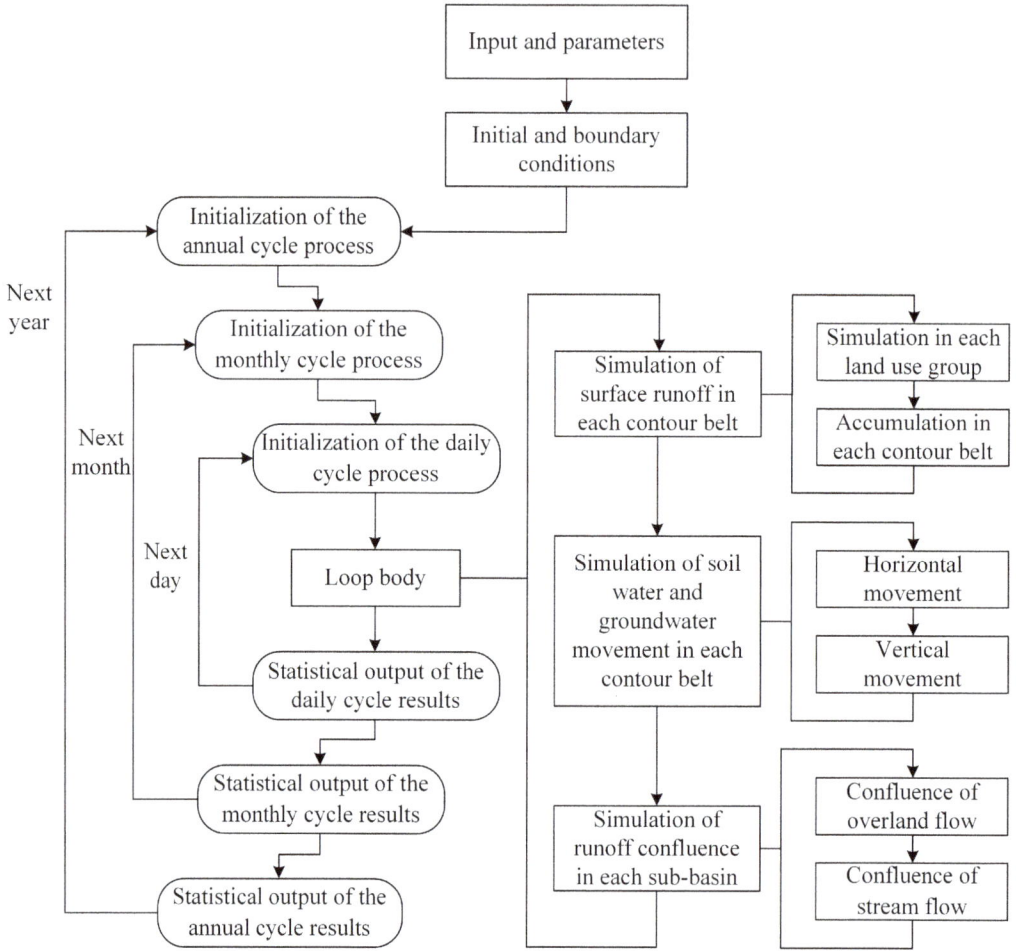

FIGURE 1.10   Calculation flowchart of the CWAM.

the region varies significantly linearly with elevation, and the spatial three-dimensional spreading method considering the influence of elevation is used to complete the interpolated spreading of the meteorological data on these subzones. Instead, a planar interpolation algorithm is used. Finally, the daily meteorological data are spread to all modeled sub-basins.

1.4.2 Zonal Calculation of Monthly Vegetation Data

Vegetation cover (*veg*) and leaf area index (*LAI*) are the two vegetation parameters that most significantly affect vegetation canopy interception, evapotranspiration, and transpiration and are characterized by seasonal and vertical zonal changes. First, the MODIS dataset of *LAI* and the *veg* data extracted from NDVI using the image meta-dichotomous model are used as data sources for the model LAI and veg. Furthermore, the large-scale area is divided into subzones, taking into account elevation changes. Finally, based on the land use type data of each subzone, its 12-month LAI and veg are counted according to the classification of forests, grasslands, and crops to form a database of vegetation parameters.

1.4.3 Identification of the Structure Type of Vadose Zones in Each Sub-Basin

The study classified the structure types of vadose zones that may occur in China into four types: general soil bedrock type, loess swelling type, soil-epikarst type, and permafrost type. Therefore, the identification of the structure type of vadose zones to which the sub-basin belongs is the basis for correcting the water movement parameters of vadose zones. First, by analyzing and combining the profile of the study area to determine whether there is a wide range of special structures of vadose zones other than the soil bedrock type in the country. Then, the extent of the area with a particular vadose structure is determined with the help of the results of previous research or through methods such as cluster analysis. Finally, according to the position where the center of mass of sub-basin is located to determine the type of vadose structure it belongs to, set the parameter ID_Sub(ix) to identify it in order to call modules 35–37, where ix is the code of sub-basin. When ID_Sub is 0, 1, 2, and 3, it indicates that the vadose structural types of the sub-basin are general bedrock type, loess swelling type, soil-epikarst type, and permafrost type, respectively.

1.4.4 Improved Simulation of Soil Water Movement for Different Structures of Vadose Zones

On the one hand, the special structures of loess swelling type, soil-epikarst type, and permafrost type were analyzed in terms of the mechanism of their influence on soil water movement, and their corresponding mathematical simulation methods were proposed. For the loess swelling type area, the swelling soil deformation when absorbing water is mainly affected by the soil swelling pressure and self-weight stress. The soil swelling pressure varies with soil water content, and the self-weight stress varies with soil depth. This caused the decrease of porosity in the soil profile, which greatly affects the soil saturated moisture movement parameters and infiltration process. The soil-epikarst type area has a particular structure that includes soil, epikarst, and unsaturated zones. The uppermost soil layer is usually thin and sometimes even missing. Epikarst zone plays a key role in water storage and movement, as well as in transpiration of vegetation root systems, where the fissure network is strongly developed. The unsaturated zone has a poor fissure development, and its water capacity and permeability are significantly lower compared to those in the epikarst zone. For the permafrost-type area, the permafrost layer also acts as a barrier to water and affects the water cycle process. Furthermore, the changes in the depth of soil freezing and thawing play an important regulatory role in runoff in permafrost areas. Therefore, the study focused on quantitatively analyzing the variation of soil saturated water content and saturated hydraulic conductivity with soil depth in loess swelling type area, the role of epikarst zone on soil water storage and hydraulic conductivity in karst-developed areas, as well as the water and heat coupling process of permafrost in cold regions. The three parts are elaborated in detail by selecting typical basins in Chapters 3–5, respectively. On the other hand, the occurrence of air resistance in soils is common in rainfall conditions.

Based on the Green–Ampt model and the results of laboratory infiltration-runoff experiments, a modified Green–Ampt model was proposed to simulate infiltrations into layered soil profiles with the entrapped air under unsteady rainfall conditions, and introduced to the CWAM. Details are set out in Chapter 6.

1.4.5 Partition Settings for Model Parameters

Firstly, the large-scale region is divided into a number of parameter subzones based on climatic conditions, geological and geomorphological differences, and water distribution characteristics. Then, according to the type of vadose structure in each subzone, set the corresponding parameter type and its initial value. General parameters include maximum depression storage depth of land surface, soil saturated hydraulic conductivity, permeability of riverbed material, Manning coefficient, snow melting coefficient, and critical air temperature for snow melting. In addition, it is necessary to set several different parameters in special types of areas, such as loess swelling type, soil-epikarst type, and permafrost type.

REFERENCES

Abbott, M.B., Bathurst, J.C., Cunge, J.A. et al., 1986a. An introduction to the European hydrological system—Systeme Hydrologique Europeen, "SHE", 1: History and philosophy of a physically-based, distributed modelling system. Journal of hydrology, 87(1–2), 45–59.

Abbott, M.B., Bathurst, J.C., Cunge, J.A. et al., 1986b. An introduction to the European hydrological system—Systeme hydrologique European, "SHE", 2: Structure of a physically-based, distributed modelling system. Journal of hydrology, 87(1–2), 61–77.

Arnold, J.G. and Fohrer, N., 2005. SWAT2000: Current capabilities and research opportunities in applied watershed modelling[J]. Hydrological processes, 19(3), 563–572.

Avissar, R. and Pielke, R.A., 1989. A parameterization of heterogeneous land-surface for atmospheric numerical models and its impact on regional meteorology. Monthly weather review, 117, 2113–2136.

Brutsaert, W., 1982. Evaporation into the Atmosphere: Theory, History, and Applications. Kluwer academic, Dordrecht.

Calver, A., 1988. Calibration, sensitivity and validation of a physically-based rainfall-runoff model. Journal of hydrology, 103(1–2), 103–115.

Chen, R.S., Kang, E.S., Wu, L.Z. et al., 2005. Cold regions in China. Cold regions science and technology, 27(4), 469–475. (in Chinese).

Ciarapica, L. and Todini, E., 2002. TOPKAPI: A model for the representation of the rainfall-Runoff process at different scales. Hydrological processes, 16(2), 207–229.

Hartmann, A., Goldscheider, N., Wagener, T. et al., 2014. Karst water resources in a changing world: Review of hydrological modeling approaches. Reviews of geophysics, 52(3), 218–242.

Hewlett, J.D., 1982. Principles of Forest Hydrology. University of Georgia Press, Athens, GA.

Hu, Z. and Islam, S., 1995. Prediction of ground surface temperature and soil moisture content by the force-Restore method. Water resources research, 31(10), 2531–2539.

Jenson, S.K. and Domingue, J.O., 1988. Extracting topographic structure from digital elevation data for geographic information system analysis. Photogrammetric engineering and remote sensing, 54(11), 1593–1600.

Jia, Y., Kinouchi, T. and Yoshitani, J., 2005. Distributed hydrologic modeling in a partially urbanized agricultural watershed using WEP model. Journal of hydrologic engineering, 10(4), 253–263.

Jia, Y., Ni, G., Kawahara, Y. et al., 2001. Development of WEP model and its application to an urban watershed. Hydrological processes, 15(11), 2175–2194.

Jia, Y., Ni, G., Yoshitani, J. et al., 2002. Coupling simulation of water and energy budgets and analysis of urban development impact. Journal of hydrologic engineering, 7(4), 302–311.

Jia, Y. and Tamai, N., 1997. Modeling infiltration into a multi-layered soil during an unsteady rain. Journal of hydroscience and hydraulic engineering, 16(2), 1–10.

Jia, Y. and Tamai, N., 1998. Integrated analysis of water and heat balance in Tokyo metropolis with a distributed model. Journal of Japan society of hydrology and water resources, 11(1), 150–163.

Jia, Y., Wang, H., Zhou, Z. et al., 2006. Development of the WEP-L distributed hydrological model and dynamic assessment of water resources in the Yellow River basin. Journal of hydrology, 331(3–4), 606–629.

Kirkby, M.J. and Beven, K.J., 1979. A physically based, variable contributing area model of basin hydrology. Hydrological sciences journal, 24(1), 43–69.

Lee, T.J. and Pielke, R.A., 1992. Estimating the soil surface specific humidity. Journal of applied meteorology and climatology, 31, 480–484.

Li, X., Cheng, G.D. and Jin, H.J., 2008. Cryospheric change in China. Global and planetary change, 62, 210–218.

Liang, X., Lettenmaier, D.P., Wood, E.F. et al., 1994. A simple hydrologically based model of land surface water and energy fluxes for general circulation models. Journal of geophysical research: Atmospheres, 99(D7), 14415–14428.

Liu, H., Du, J., Jia, Y. et al., 2019. Improvement of watershed subdivision method for large scale regional distributed hydrology mode. Advanced engineering sciences, 51(1), 36–44. (in Chinese)

Maidment, D.R., 1992. Handbook of Hydrology. McGraw-Hill, New York.

Mein, R.G. and Larson, C.L., 1973. Modelling infiltration during a steady rain. Water resources research, 9(2), 384–394.

Monteith, J.L., 1973. Principles of Environmental Physics. Edward Arnold Publishers, London.

Moore, I.D. and Eigel, J.D., 1981. Infiltration into two-layered soil profiles. Transactions ASAE, 24, 1496–1503.

Noilhan, J. and Planton, S.A., 1989. Simple parameterization of land surface processes for meteorological models. Monthly weather review, 117, 536–549.

Perrin, J., Jeannin, P.Y. and Zwahlen, F., 2003. Epikarst storage in a karst aquifer: A conceptual model based on isotopic data, Milandre test site. Switzerland. Journal of hydrology, 279(1–4), 106–124.

Rui, X.F., 1996. Discussion of some problems on mechanism of runoff yield. Journal of hydraulic engineering, 9, 22–26. (in Chinese).

Sevruk, B., 1997. Regional dependency of precipitation-altitude relationship in the Swiss Alps. Climatic change, 36(3–4), 355–369.

Siriwardena, L., Finlayson, B.L. and Mcmahon, T.A., 2006. The impact of land use change on catchment hydrology in large catchments: The Comet River, Central Queensland, Australia. Journal of hydrology, 326(1–4), 199–214.

Su, N., 2010. Theory of infiltration: Infiltration into swelling soils in a material coordinate. Journal of hydrology, 395, 103–108.

Verdin, K.L. and Verdin, J.P., 1999. A topological system for delineation and codification of the Earth's river basins. Journal of hydrology, 218, 1–12.

Wagener, T., Wheater, H.S. and Gupta, H.V., 2004. Rainfall-Runoff Modelling in Gauged and Ungauged Catchments. Imperial College Press, London.

Wang, L., Wang, Z.J., Yin, H. et al., 2006. A distributed hydrological Model-GBHNM and its application in middle-scale catchment. Journal of glaciology and geocryology, 28(2), 256–261.

Wang, G.Q., Zhang, J.Y., He, R.M. et al., 2007. Trends of temperature change in middle of Yellow River and its impact to the evaporation potential. Journal of water resources and water engineering, 18(4), 32–36. (in Chinese).

Watanabe, K. and Osada, Y., 2017. Simultaneous measurement of unfrozen water content and hydraulic conductivity of partially frozen soil near 0 °C. Cold regions science and technology, 142, 79–84.

Xia, J., Wang, G.S., Tan, G. et al., 2004. Distributed time-variant gain hydrological model. Scientia sinica (Terrae), 34(11), 1062–1071.

Zhang, Z.C., Chen, X., Ghadouani, A. et al., 2011. Modeling hydrological processes influenced by soil, rock and vegetation in a small Karst basin of southwest China. Hydrological processes, 25(15), 2456–2470.

Zhang, Z.L., Fan, H.M., Guo, C.J. et al., 2008. Study on the relationship between flow velocity of sloping face and runoff characteristic under the simulated rainfall. Research of soil and water conservation, 15(6), 32–34. (in Chinese).

CHAPTER 2

New Sub-Basin Division Method for Large-Scale Regions and Its Application in China

2.1 RESEARCH BACKGROUND

In recent years, distributed hydrological modeling, which takes into account the spatial heterogeneity of hydrological parameters and processes, has become an important tool for basin/regional hydrological studies. Distributed hydrological models need to discretize the study area into many smaller spatial units and assume their internal homogeneity. As a result, the hydrological simulation of these small units can be used as the basis for the simulation of water cycle processes in the whole study area. Currently, distributed hydrological models are mainly used to achieve spatial division of the study area in the form of rasters and sub-basins. When the spatial scale of the study area is large, the latter discretization is often used to take into account the accuracy and efficiency of simulation. Consequently, the study area is divided into a certain number of sub-basins, which are used as the basic simulation units. Therefore, reasonable sub-basin division and accurate portrayal of the study area scope are crucial for regional/basin-distributed hydrological modeling and water resources assessment.

Currently, the commonly used sub-basin division methods based on surface runoff and flow model are mostly aimed at complete basins. The related research focuses on the innovations of digital river network extraction algorithms (Barnes et al., 2014) and sub-basin coding methods (Liu et al., 2014), the effects of different DEM data sources on sub-basin division (Sousa and Paz, 2017), and the determination of the suitable catchment area thresholds (Reddy et al., 2018). However, there are few distributed modeling studies for large-scale regions. The sub-basin division for large-scale regions mainly has the following

DOI: 10.1201/9781003646648-2

characteristics: (i) the study area is not a complete basin, which does not have the characteristic of "closure", and the number of outlets in the boundary is not fixed; (ii) the rivers are numerous and independent of each other, and the characteristics of river systems vary considerably, with the co-existence of inland rivers and outflow rivers; (iii) complex topography and geomorphology lead to great variability in the spatial distribution of rivers; and (iv) the existence of micro-topographic areas with drastic changes in elevation results in the huge difference in catchment area between basins with no confluence connection in the study area.

The traditional sub-basin division method has obvious shortcomings in large-scale regions. First, the method has operational difficulties in determining the outlet points, which require a lot of manual positioning work. Second, due to the limitations of manual operation, some key outlet points are easy to miss. A more prominent contradiction is reflected in the threshold setting, where a single catchment threshold is difficult to balance the model operation efficiency and simulation accuracy. Too small a threshold will lead to a surge in computation, while too large a threshold will reduce the spatial resolution. This technical limitation directly affects the optimized application of hydrological models.

Therefore, for large-scale regional distributed simulation, based on the DEM river network stem and branch topology coding rules, this chapter proposes a sub-basin division method, which is based on automatic identification of basin outlet and multi-threshold virtual river network fusion technology. The method controls the number of sub-basins through accurately grasping the scope of the study area, takes into account the model calculation accuracy and efficiency, and provides support for large-scale regional distributed modeling. Finally, China is chosen as a research example to complete the sub-basin division and unified coding in the construction of the China Water Assessment Model (CWAM).

2.2 LIMITATIONS OF TRADITIONAL SUB-BASIN DIVISION METHODS APPLIED IN CHINA

2.2.1 Distributed Hydrological Model and Sub-Basin Division

Depending on the geographical characteristics of the study area, distributed hydrological models can be divided into two categories: river basin models and regional models. Regardless of the model type, distributed hydrological modeling involves discretizing the study area into a number of smaller spatial units. The hydrological simulation of these units serves as the basis for the simulation of the water cycle processes in the whole study area. Currently, distributed hydrological models achieve spatial discretization of the study area mainly through two forms: raster and sub-basin. When the spatial scale of the study area is large, the latter discrete approach is frequently employed to balance simulation accuracy and computational efficiency. Specifically, the study area is divided into a certain number of sub-basins, which serve as the basic computation units. Consequently, the sub-basin division is a critical foundational task for distributed hydrological modeling in large-scale regions, directly influencing the precision of hydrological simulations and forecasts. With the development of the distributed hydrological model, higher targets are required for the effectiveness of sub-basin division.

2.2.1.1 Accurate Extraction of the Virtual River Network

River network characteristics are a comprehensive reflection of the hydrological characteristics of the study area, and the accuracy of river network extraction is an important factor that restricts the simulation effect of the model. Nowadays, a widely used method for river network extraction is the surface overland flow modeling method. However, it suffers from flow path distortion in depressions and flat areas. The International Hydrological Decade (IHD) has designated "Panta Rhei" as the only development direction for 2013–2022, and distributed hydrological modeling is gradually expanding from mountainous areas to areas with high anthropogenic impacts. However, compared with mountainous areas, the characteristics of river networks under intense human impacts become more complex, and the extraction accuracy faces more challenges.

2.2.1.2 Appropriate Size of Sub-Basins

The size of sub-basins reflects the spatial discretization of the study area by the distributed hydrological model and directly affects runoff generation and confluence processes in the study area. Specifically, if the sub-basins are small, a large number of redundant sub-basins will be generated and significantly increase the computation burden of the model. If the sub-basins are large, they fail to reflect the inhomogeneity of the subsurface in the study area, which is not conducive to the refined simulation of the model. In addition, many natural or artificial rivers are very long, with no tributaries joining in the middle, eventually dividing into several narrow sub-basins, which affects the efficiency of the model calculations. Therefore, the extracted river network needs to be further processed to obtain suitable sizes of sub-basins to satisfy modeling needs.

2.2.1.3 Efficient Coding of Sub-Basins

Coding of sub-basins is a prerequisite for identifying runoff confluence pathways and carrying out modeling of hydrological processes. The more complex the river network in the study area, the higher the coding requirements. Specifically, the codes should correspond to the river sections, which can reflect the upstream and downstream relationship of the sub-basins. If the coding rules of the sub-basins are known, the code of the upstream and downstream sub-basins can be calculated directly without the need for tedious traversal. In addition, the extraction of the river network by DEM will result in multiple river sections converging into the same reach, as well as generating several long, narrow river sections, placing greater demands on sub-basins coding.

2.2.2 Traditional Sub-Basin Division Methods

Traditional sub-basin division is commonly based on the DEM-based overland flow modeling method proposed by Jenson and Domingue (1988). The method mainly includes the extraction of the virtual river network, the determination of the outlet and split points of the basin, and the division and coding of the sub-basins, as shown in Figure 2.1. Resampling the original DEM raster to an appropriate resolution is the first step in extracting the virtual river network. Then, the depressions in the resampled DEM raster are filled. Furthermore, the measured river network is introduced as a constraint to correct the raster

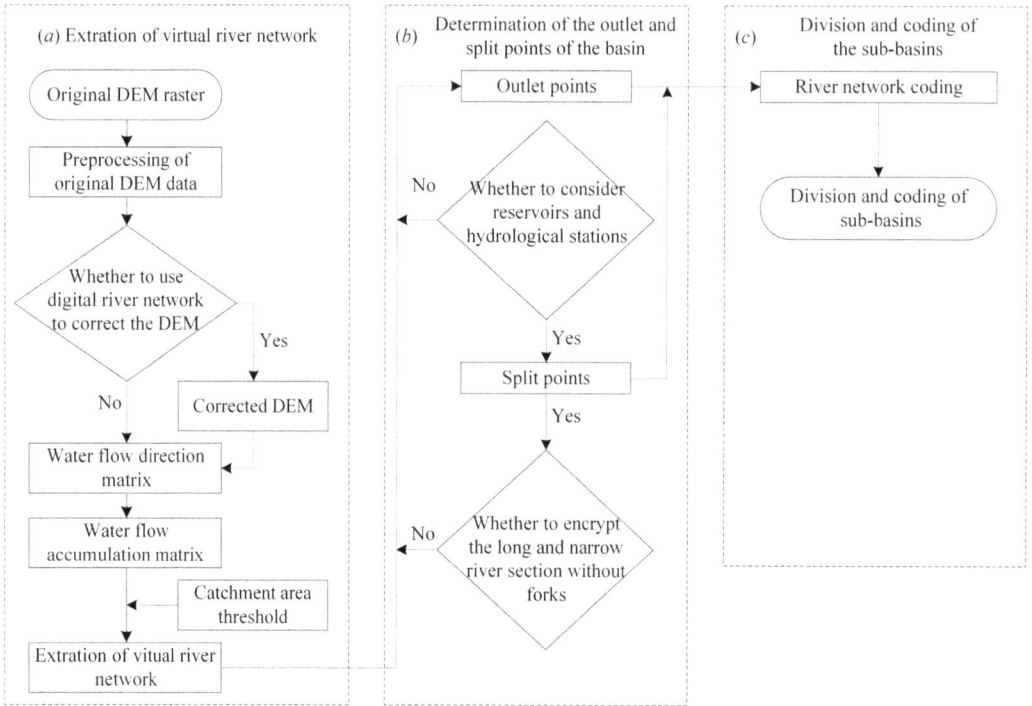

FIGURE 2.1   Flow chart of traditional sub-basin division methods.

to enhance the flow path information. The water flow direction matrix and water flow accumulation matrix are calculated, and a suitable catchment area threshold is determined to extract the virtual river network.

To improve the extraction effect of river networks, scholars at home and abroad have done a lot of research work on DEM raster scale conversion, DEM data correction, flow direction algorithm improvement, and catchment area threshold determination. When extracting the virtual river network for sub-basin division, it is necessary to manually set the outlet point and split points of the river basin. In this case, the basin outlet point is always located at the edge of the study area. The split points are made by considering the location of reservoirs and hydrological stations, as well as the encryption needs of narrow sub-basins. Finally, the divided sub-basins are coded. Topological attribute table method, bifurcation tree coding, multi-furcation tree coding, Pfafstetter method, and stem and branch topological coding are the main coding methods used. However, these methods are not yet able to fully meet the requirements of efficient sub-basin coding, especially in large-scale region.

2.2.3 Difficulties in Applying Traditional Methods to Large-Scale Regions

Traditional sub-basin division methods face three difficulties when applied to large-scale regions such as China:

2.2.3.1 Difficulty in Accurately Capturing the Spatial Extent of the Study Area

The simulation extent of the study area determines its water heat flux, which is directly related to the model simulation results. The catchment area threshold is one of the key parameters when extracting the virtual river network and determining the modeling extent of the study area. For a comprehensive river basin analysis, an optimal catchment area threshold must be established. DEM rasters exhibiting water flow accumulations exceeding this threshold are then classified as river network rasters, thereby constructing a virtual river network. Consequently, a higher threshold results in a sparser extracted virtual river network and a coarser division of the sub-basins. Conversely, the denser the virtual river network, the correspondingly higher the number of sub-basins to be divided, and the higher the considerations of spatial variability in the study area. Therefore, there is a need to determine an optimal catchment area threshold that accurately reflects the characteristics of the river network and also controls the density of the network. If the river network can be accurately extracted, then the modeling extent can be effectively controlled and can be in good agreement with the actual basin boundary. However, the sub-basin division by determining a single catchment area threshold is not applicable in large-scale complex terrain areas like China. The reason is that large-scale complex terrain areas contain multiple intact and non-intact basins that are independent of each other and have no confluence but have widely varying catchment sizes. In this case, a single catchment area threshold cannot reconcile the dual needs of model computation efficiency and modeling extent accuracy. Specifically, the catchment area threshold needs to be set very small if it is to ensure that basins with small catchment areas in micro-topographic areas can be extracted. However, for distributed hydrological modeling in large-scale areas, the large number of sub-basins will lead the model to a computational disaster, and the simulation efficiency will be reduced significantly. On the contrary, if the computation efficiency of the model is considered, the catchment area threshold needs to be increased and the number of sub-basins needs to be reduced. In this case, many independent river networks with small catchment areas in micro-topographic areas will not be able to be extracted, resulting in the loss of part of the simulated area.

2.2.3.2 Difficulty in Accurately Identifying the Basin Outlet

In traditional sub-basin division methods, the virtual river network grid located on the boundary of the study area is generally set up artificially as outlets. For complete basins, the basin outlet is unique and easy to determine. However, large-scale complex terrain areas are confronted with a large number of rivers and may involve transboundary rivers, resulting in a large number of basin outlets. In this case, manually setting the basin outlets is inefficient and prone to omission. Additionally, pseudo-outlets may exist along the study area boundary, such as outlet points 1 and 2 in Figure 2.2.

2.2.3.3 Difficulty in Accurately Reflecting Inland Rivers

Before extracting the virtual river network using the water flow accumulation method, the resampled DEM rasters need to be pre-processed by filling in the depressions to solve the problem of "digital negative terrain". The basic idea of this measure is to first find the depressions that exist in the DEM raster and then fill them in by gradually padding them. This allows all DEM rasters in the river basin to eventually flow out of the outlet. It can be argued that this approach applies only to outflow rivers entering the ocean. However, in large-scale complex terrain areas where inland rivers and outflow rivers co-exist, the

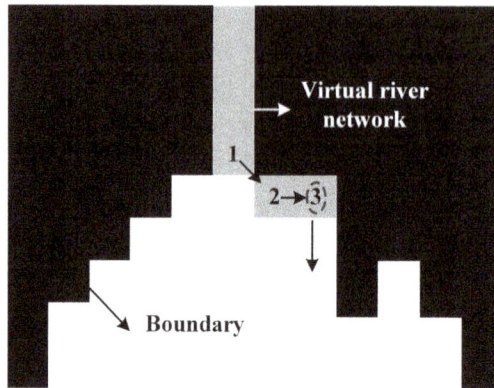

FIGURE 2.2 Schematic distribution of outlets in a river basin.

traditional river network extraction process will not accurately reflect the true status of inland rivers. Specifically, inland rivers tend to develop within closed basins and eventually disperse mostly in downstream irrigation areas or deserts, with a few forming tailrace lakes in lowlands. Traditional methods for inland rivers are still looking for potential outlets along the study area boundary when filling puddles on the DEM raster. This method ultimately makes the extracted virtual river network flow out of the boundary, which is seriously inconsistent with the actual inland river network, as shown in Figure 2.3.

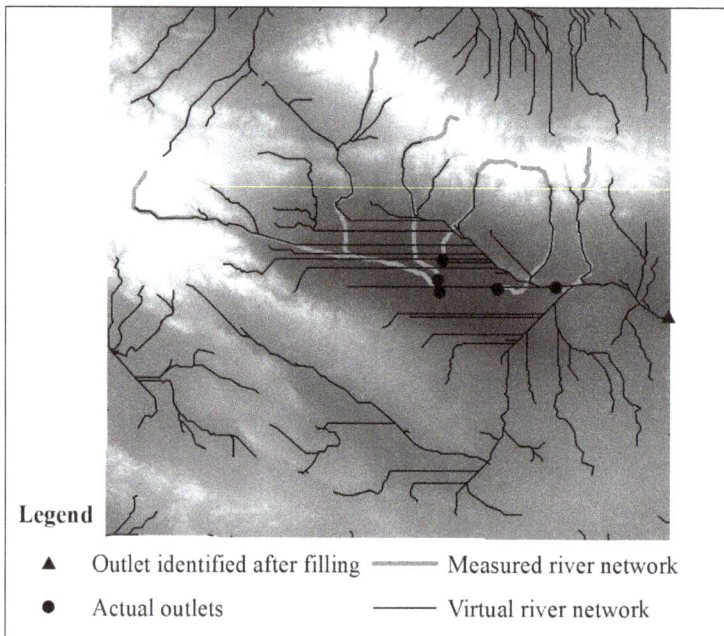

FIGURE 2.3 Schematic diagram of virtual river network extraction for an inland river.

2.3 CALCULATION PROCESS FOR NEW SUB-BASIN DIVISION METHOD

2.3.1 Overall Process Design of the Method

Aiming at the above difficulties, the study proposed a new sub-basin division method. The method was based on automatic recognition of the outlets of basins and the fusion of the river network with variable catchment area thresholds. It consists of four major steps: (*a*) automatic recognition and determination of basin outlets; (*b*) threshold determination of large and small catchment areas; (*c*) multi-thresholds virtual river network fusion; and (*d*) sub-basin division and coding. The calculation process is shown in Figure 2.4.

2.3.2 Automatic Recognition and Determination of Basin Outlets

The study identified and determined basin outlets for two types of rivers, namely, inland rivers and outflow rivers entering the ocean. For inland river outlets, considering that the rivers are relatively sparse, their outlets were mainly determined manually. Specifically, the digital layer of measured inland rivers was drained, and the DEM raster elevation value at the end of the river was set to null. Furthermore, virtual river network extraction was completed by finding outlets of inland rivers around the null raster using traditional methods.

For outflow rivers entering the ocean, the study innovated a method for automatic recognition and determination of basin outlets. The process was as follows: ① Using the measured river network, the effective information of the DEM raster was increased by

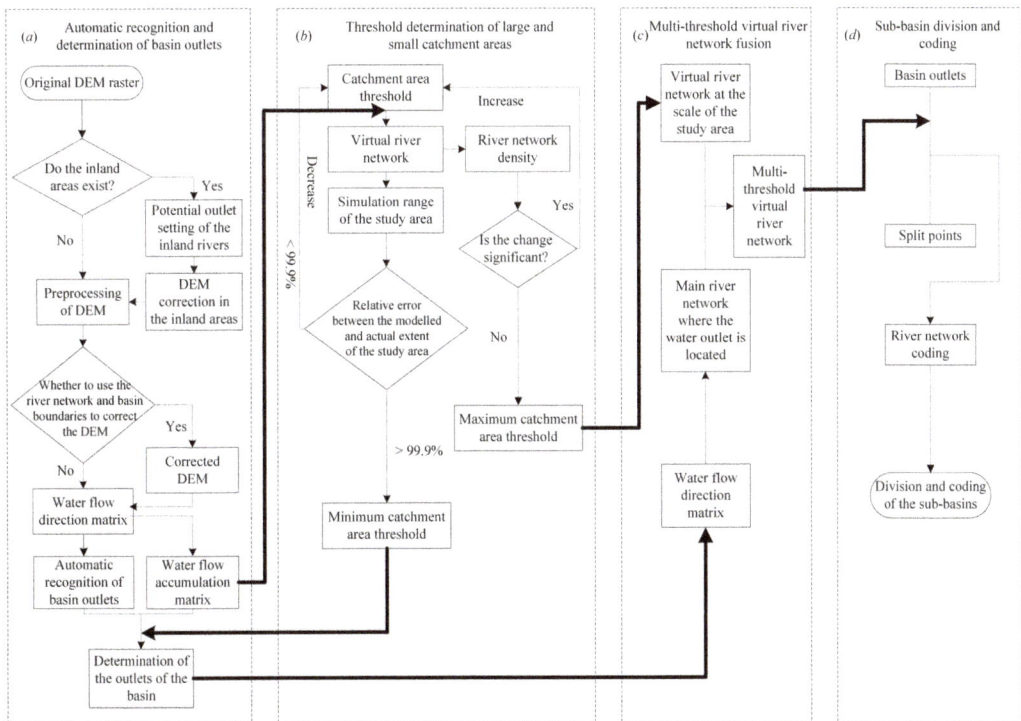

FIGURE 2.4   A method based on automatic recognition of the outlets of basins and the fusion of the river network with variable catchment area thresholds.

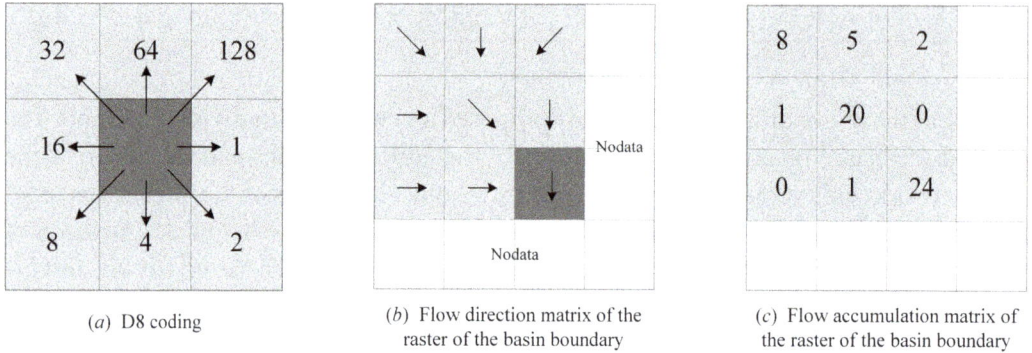

(a) D8 coding

(b) Flow direction matrix of the raster of the basin boundary

(c) Flow accumulation matrix of the raster of the basin boundary

FIGURE 2.5 Flow direction and cumulative volume calculations for raster flows at boundaries (D8 algorithm).

river burns to generate rasters for flow direction and accumulation calculations, denoted as "dem_burn"; ② Based on the raster "dem_burn", the flow direction raster "flowdir" and the flow accumulation raster "flowacc" were calculated for the study area; ③ Following the D8 flow direction algorithm, the outlet of outflow river is located within a 3×3 raster window centered on the basin boundary raster. Therefore, the vector boundary line of the study area was buffered by 3 rasters to generate a vector layer, notated as "outline". As a result, the file "outlet" completely covers all the outlets of outflow rivers in the study area; ④ Iterates over the flow directions of rasters within the file "outline". For a given raster, if the flow direction value of the next raster to which it flows is null, the raster is marked as an "alternative" basin outlet, as shown in the dark raster in Figure 2.5(b); ⑤ Determine the small catchment area threshold for the study area (see Section 2.2.3 for details), and based on this threshold determine the final basin outlet from the "alternative" outlets.

2.3.3 Multi-thresholds Virtual River Network Extraction

Aiming at the difficulties encountered in applying a single catchment area threshold to sub-basin division in large-scale complex terrain areas, a multi-thresholds river network extraction and fusion method was proposed. Through this method, the extracted river network can be made to accurately reflect the river characteristics of the study area. Meanwhile, the density of the regional river network can be controlled within a reasonable range to realize the control of the number of sub-basins.

2.3.3.1 Determination of Large Catchment Area Thresholds for River Network Extraction

The river network density method was used to determine a threshold suitable for macroscopic characterization of river systems in the study area as the large catchment area threshold. Using this catchment area threshold, most of the sub-basins within the study area can be accurately extracted. The so-called river network density method is used to generate a series of virtual river networks by selecting different catchment area thresholds. Then, the relationship curve of the density of virtual river networks with the catchment area threshold is plotted. As the threshold value increases, the virtual river network density tends to level off after a sudden drop. The point where the virtual river network density no longer changes significantly with the increase of the threshold value was identified to be the optimal catchment area threshold. This threshold was determined to be the large catchment area threshold in this study.

2.3.3.2 Determination of Small Catchment Area Thresholds for River Network Extraction

The large catchment area threshold does not handle micro-topographic areas with high elevation relief near the study area boundary. Therefore, a small catchment area threshold needs to be determined that can be used to accurately characterize the extent of the study area. Regarding the modeling accuracy of the extent of the study area, it is mostly measured by the relative error between the modeled and actual extent of the study area, but there is no fixed standard. In this study, an absolute value of relative error of 0.1% was chosen as the criterion for the simulation accuracy to meet the standard. That is, the study area can be accurately described when the agreement between the simulated and actual ranges of the study area is 99.9% and above. Based on the above ideas, the catchment area threshold was continuously reduced, and sub-basins were divided. When the degree of agreement reaches 99.9%, the corresponding threshold was selected as the small catchment area threshold in this study.

2.3.3.3 Generalization and Fusion of Virtual River Networks for Multi-Thresholds Extraction

When the small catchment area threshold is chosen, a large number of narrow, redundant sub-basins appear. These sub-basins contribute little to large-scale regional distributed hydrological modeling but significantly increase model computation. To address the above issues, the virtual river network extracted from the small catchment area threshold is further generalized. The principle of generalization is as follows: take the raster where the outlet of the micro-topographic area catchment is located as the starting raster, and identify the upstream rasters of this raster through the water flow direction. The upstream rasters can be one or more. If there is only one upstream raster, this upstream river raster is assigned a value of 1 as the mainstem raster directly. If there are more than one upstream raster (e.g., raster A and B converge into the river network grid ① at the same time in Figure 2.6), compare the size of the cumulative number of water flows of these rasters, and select the raster with the largest cumulative number (A) as the mainstem raster, assigning the value of 1. Furthermore, the mainstem raster was selected as the starting raster to continue tracing upstream the mainstem raster to the river headwaters. As a result, a mainstem stream network can be traced based on each basin outlet in the micro-topographic area (e.g., thick dark line in Figure 2.6). With this measure, the cumulative number of water flows at the outlet of each catchment in the micro-topographic area is the threshold for virtual river network extraction in the catchment. These thresholds vary in size between large and small catchment area thresholds. Finally, the mainstem river network generalized for micro-topographic areas was fused with the virtual river network obtained under the large catchment area threshold.

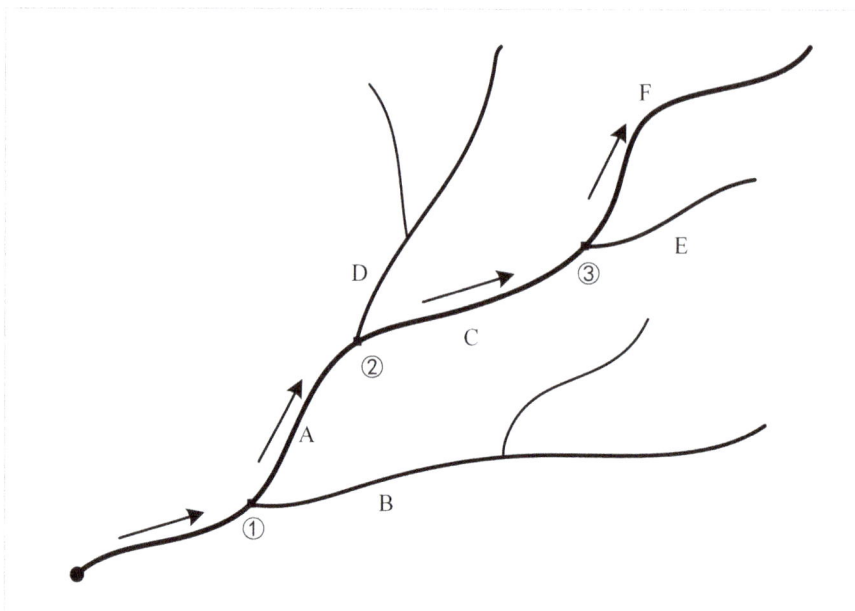

FIGURE 2.6 Schematic diagram of virtual river network generalization for small catchment area threshold.

2.3.4 Sub-basin Division and Coding

Considering the location of reservoirs and hydrological stations, as well as the need for narrow river segmentation, sub-basin division was carried out by setting up splitting points on the virtual river network. Furthermore, the stem-branch topological codification (SBTC) method was developed to code the sub-basins (Liu et al., 2014). A more robust coding system is powerful for the effective application of complex hydrological models to large-scale river networks. In this method, each reach has only one code, and each code indicates only one reach, and the codes should reflect the topological relationship of the reaches in the virtual river networks. In short, the basic idea of SBTC is to consider a river as a topology consisting of a mainstem with several tributaries that feed into it. If the tributaries are continued, the tributaries are also composed of the mainstem and its tributaries. For a given river basin, the reach with the largest catchment area is selected as the mainstem reach by comparing the catchment area. After the mainstem reaches are identified, the reaches that feed into the mainstem are considered tributary reaches. In addition, these reaches would be further divided into multiple reaches by split points. Considering the large number of rivers in China, the coding system consisted of four parts: outlet code, mainstem reach code, tributary reach code, and upstream sub-basin number code. The system was characterized by {O, S, B: U}, as shown in Figure 2.7.

First, the basin outlets in the study area were coded in order from 1 to n, where n is the total number of outlets. For Class I rivers, the mainstem reaches passed through were coded 1, 2, …, m in order, following a particular outlet back upstream to the headwater reach. This type of code is the mainstream reach code. Meanwhile, the Class II rivers that

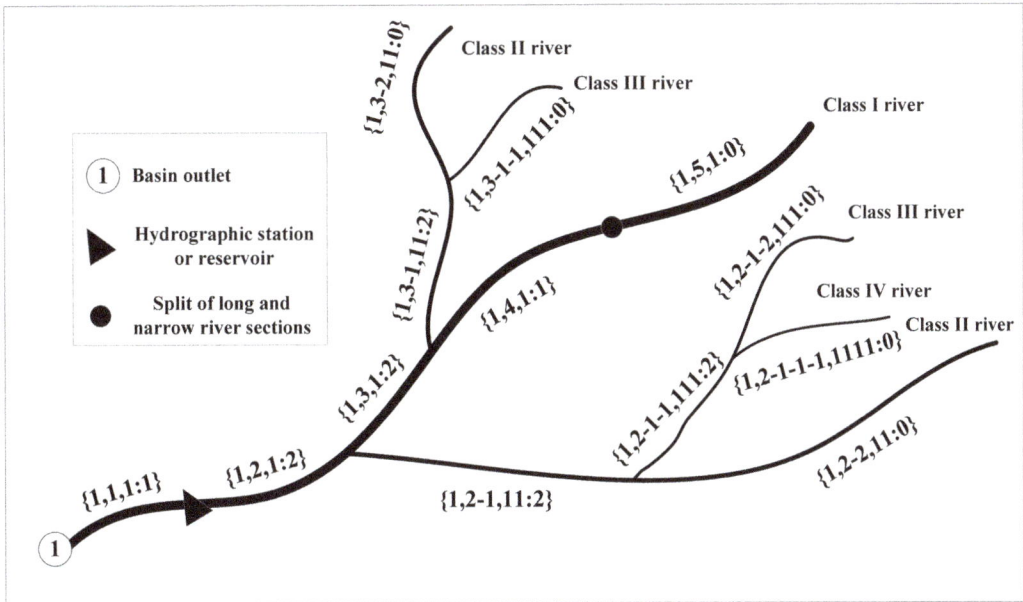

FIGURE 2.7 Schematic diagram of the sub-basin coding based on the stem-branch topological relationship of reaches.

feed into Class I rivers have mainstem reaches of their own. As a result, the mainstem code of these Class II rivers is made up of the mainstem code of the higher-level reach it feeds into and the mainstem code of its level. By this rule, the mainstem codes of the lower-level rivers can also be progressively determined. For tributary reach code, a Class I river is equivalent to one of its tributaries with a tributary code of 1 when viewed from the basin outlet. Class II river is the tributary of Class I river, and its tributary reach code consists of the tributary code of the Class I river in which it is located and its tributary code. Finally, while giving each reach a unique identification, an upstream sub-basin number code was added to quickly locate its neighboring upstream or downstream sub-basins.

2.4 EVALUATION OF THE EFFECTIVENESS OF THE NEW METHODOLOGY IN CHINA

2.4.1 Preparation of Basic Data

Due to the existence of depressions and flat areas in China, it is necessary to make corrections and adjustments to the DEM raster based on the actual river network information. Through these corrections, the extracted virtual river network is closer to the measured river network. Therefore, pre-processing of the DEM raster and the actual river network is required before sub-basin division is carried out.

2.4.1.1 DEM Raster Pre-processing in China

Considering the large-scale range of about 9.6 million km² in the country, the 1km × 1km DEM raster for sub-basin division was chosen for the computation efficiency and statistical convenience. The original DEM raster used in this study is derived from 90m×90m data acquired by NASA and USGS via Shuttle radar, covering the entire Chinese region. Firstly, the DEM raster spatial resolution was resampled from 90m to 1km. The geographic coordinate system was selected as WGS84, and the projection was an equal-area Albers projection. The projection parameter was set to the central meridian of 105°E, and the two standard lines of latitude were at 25° and 47°, respectively. In addition, there are many boundary rivers in China. If the boundary river is used as the boundary, a large number of outlets will be created, significantly increasing the workload. Consequently, the study buffered the boundary line outwards by a distance of 3 rasters to include the boundary river to determine the outlets. China has a zigzag coastline and many small islands. To improve computation efficiency, the original coastline was fine-tuned, and small islands (i.e., DEM patches) were excluded from sub-basin division.

2.4.1.2 Measured Digital River Network Pre-Processing

Some features of the measured digital river network can lead to errors in the virtual river network extraction process. For example, river channels and looped river networks are represented by double lines, and reservoir and lake boundaries are represented by line elements. Therefore, the study modified the river network file so that the extracted virtual river network is both close to reality and easy to calculate. The main modifications are set out below: ① Revise the two-lane channel to a single-lane channel; ② Revise the circular channel to a tree-like channel; ③ Consider the correction of the confluence routes of lakes and reservoirs; and ④ Straighten the overly curved channel appropriately.

2.4.2 Application of Methods

2.4.2.1 Identification of Potential Outlets of Inland Rivers

China's inland area is vast, accounting for about one-third of the country's land area. However, the river network in the inland area is sparse due to scant rainfall. A total of 67 potential outlets were manually set at the end of the rivers based on the measured river system in the inland area. As a result, the virtual river network in the inland area can be accurately extracted and well matched with the actual river network, as shown in Figure 2.8.

2.4.2.2 Small Catchment Area Threshold Determination

A number of catchment area thresholds were selected in the range of 10–1000 km², and the virtual river network density corresponding to each threshold was calculated. Furthermore, the variation curve of the catchment area threshold in relation to the density of the river network was calculated, as shown in Figure 2.9. It can be seen that the curve stabilizes when the area threshold is in the range of 200–300 km². Through further investigation, 240 km² was finalized as the threshold for a large catchment area on the national scale.

2.4.2.3 Large Catchment Area Threshold Determination

After determining the large catchment area threshold, the size of the catchment area threshold was continually reduced to assess how well the extent of sub-basin division matched the actual extent of the study area. It was found that when the catchment area

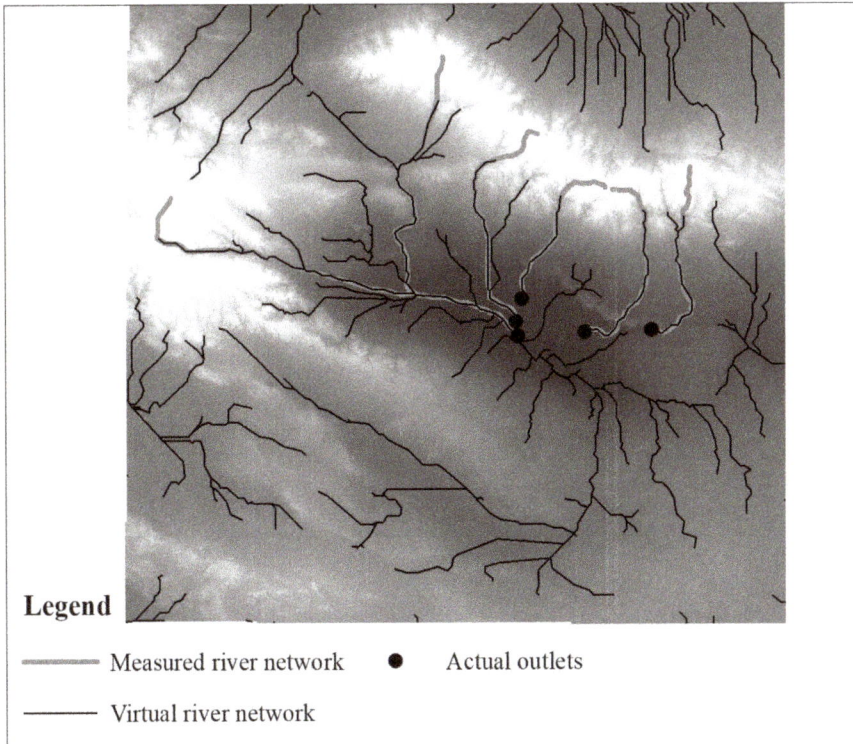

FIGURE 2.8 Schematic diagram of the effects of inland river extraction after setting up potential outlets.

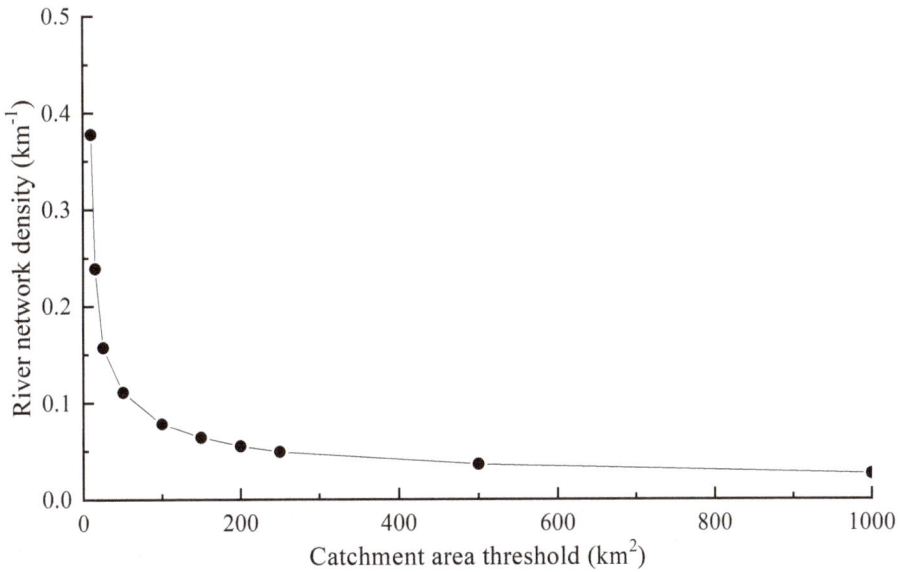

FIGURE 2.9 Catchment area versus river network density curve.

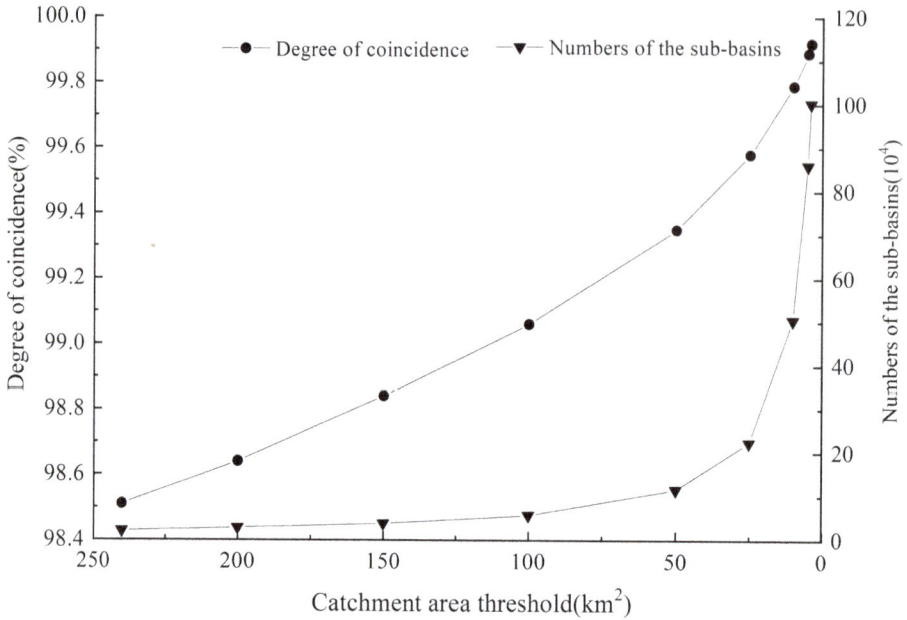

FIGURE 2.10 Curves of the degree of coincidence of sub-basin division and the number of sub-basins with catchment area thresholds.

threshold was lowered to 4 km², the simulated extent of the sub-basins coincided with the national extent by 99.92%, but the number of sub-basins increased to 100,260,000, as shown in Figure 2.10.

2.4.2.4 Determination of Basin Outlets and Split Points

First, the flow direction of the national boundary raster was determined based on the Chinese water flow direction raster data. Also, the flow direction of the raster within the perimeter of potential outlets in the inland area. With these two flow direction files, all outlets in river basins across the country were identified. Subsequently, the outlets with a water flow accumulation greater than 4 were screened, using the flow accumulation file of the raster. These outlets were identified as the final basin outlets. As a result, a total of 4121 basin outlets were identified, distributed along the coastline, borderline, and inland areas, as shown in Figure 2.11.

2.4.2.5 Convergence of Virtual River Networks with Multiple Thresholds

First, a catchment area of 240 km² was used as the extraction threshold for the virtual river network on the national scale. Next, the outlets at sub-basin area thresholds of 4 km² and 240 km² were compared. It was found that the number of outlets with the 240 km² catchment area threshold was 3406 less relative to the number of outlets with the 4 km² catchment area threshold. Consequently, the 3406 catchment outlets were used as a starting point to move upstream and trace the mainstem of the river network where the outlets

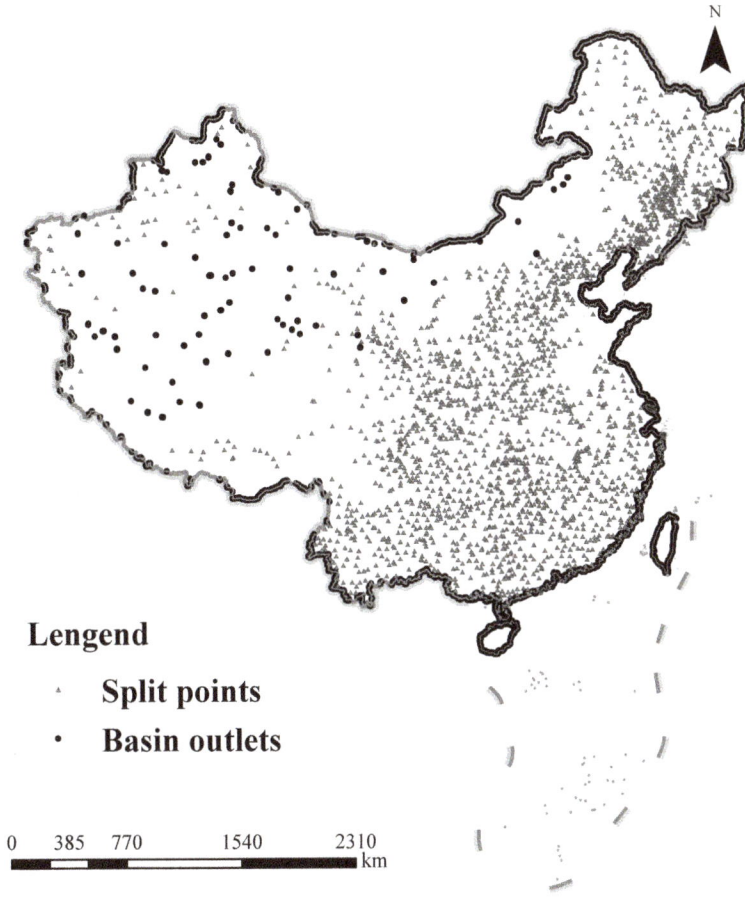

Lengend

▵ **Split points**

• **Basin outlets**

0 385 770   1540   2310
                                km

FIGURE 2.11   Distribution of basin outlets and Split points in China.

are located through the water flow direction and accumulation files. It can be determined that the catchment area threshold for these virtual river networks ranged between 4 and 240 km². Finally, the multi-threshold river network obtained from the two approaches was fused for national sub-basin division and coding. A localized illustration of Hainan Island is shown in Figure 2.12.

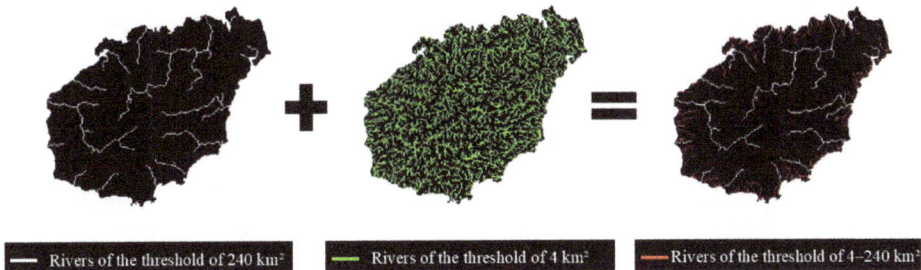

Rivers of the threshold of 240 km²   Rivers of the threshold of 4 km²   Rivers of the threshold of 4–240 km²

FIGURE 2.12   Schematic diagram of multi-threshold virtual river network integration on Hainan Island.

2.4.3 Evaluation of Application Effectiveness

The traditional method (Method I) and the new method proposed in this study (Method II) were used, respectively, to test the effectiveness of sub-basin division methods in complex terrain areas in China.

2.4.3.1 Effectiveness of River Network Extraction

Using the traditional method, with 240 km² as the catchment area threshold to extract the virtual river network, would result in many river networks with small catchment areas not being extracted. In contrast, the extraction of the river network of independent sub-catchments can be better achieved by using the new method. Meanwhile, comparing the measured river network with the results of the new method, they are in good agreement.

2.4.3.2 Effectiveness of Sub-Basin Division

Through the traditional method, a total of 21,768 sub-basins were divided within China. Of these sub-basins, the average area was 434 km², with a maximum area of 5512 km² and a minimum area of 10 km². The extent of the extracted river network matched the actual extent by only 98.51%, and sub-basin division failed in a large number of microtopographic areas along the coastline with dramatic elevation fluctuations. Through the method based on automatic recognition of the outlets of basins and the fusion of the river network with variable catchment area thresholds, 21,768 sub-basins were subdivided nationwide. The average area of these sub-basins was 381 km², with a maximum area of 5512 km² and a minimum area of 4 km². It can be seen that the new method proposed in this study is well applied in the complex terrain area in China, and the local schematic is shown in Figures 2.13 and 2.14.

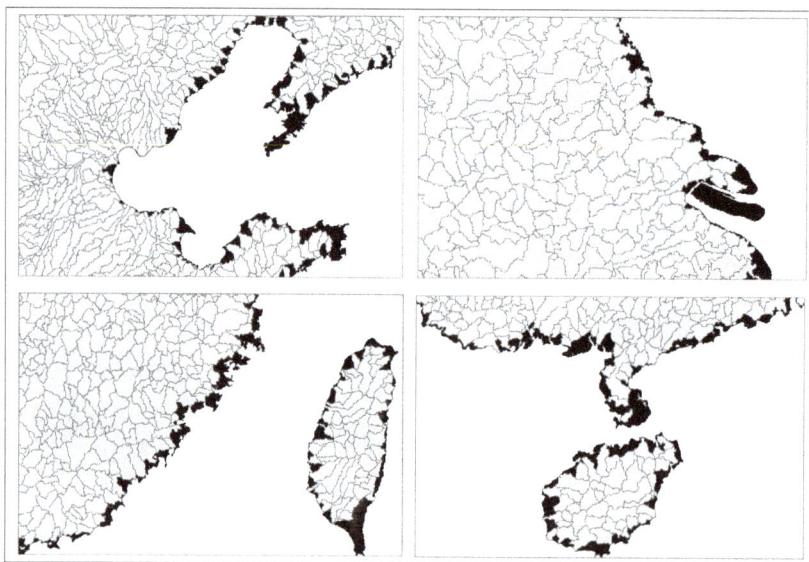

FIGURE 2.13   Schematic diagram of sub-basins in localized complex terrain areas in China based on Method I.

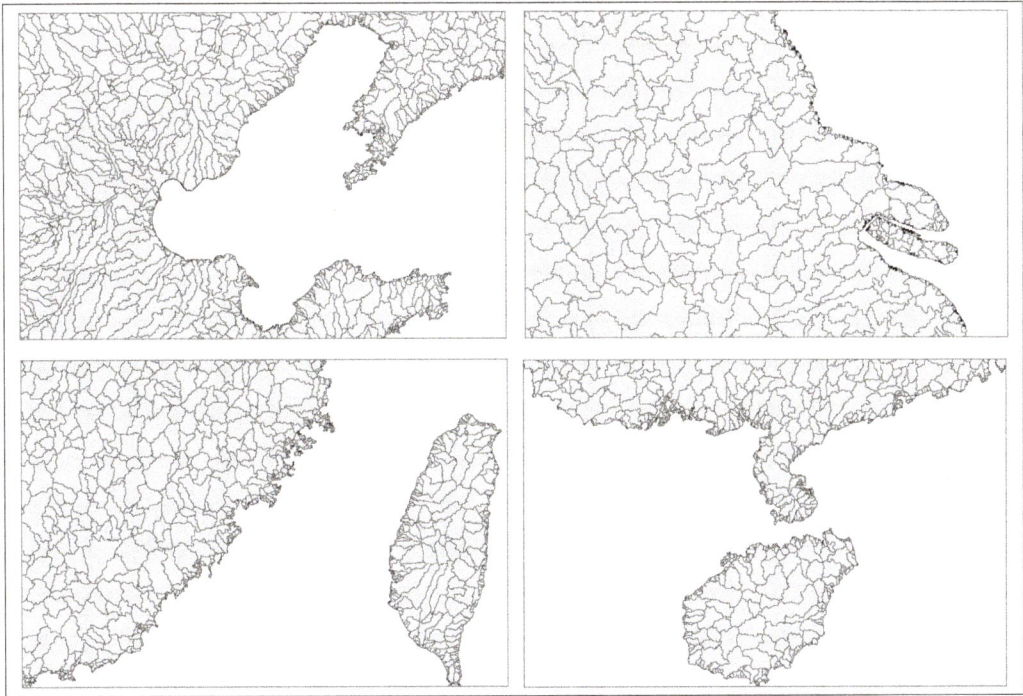

FIGURE 2.14   Schematic diagram of sub-basins in localized complex terrain areas in China based on Method II.

In addition, 10 Class I WRRs and 80 Class II WRRs in the country were selected, and the coincidence of the sub-basins' coverage with the actual extent in these regions was calculated. As can be seen from Table 2.1, the coincidence value ranged from 99.73% to 100.13%, with a deviation of no more than 0.5%. Moreover, a box plot of the change in the extent to which the sub-basins' coverage of 80 Class II WRRs coincides with its actual extent was drawn, as shown in Figure 2.15. Apart from individual Class II WRRs, the coincidence values in most of the WRRs ranged from 99% to 101%, with a mean between 99.5% and 100.5%, giving good simulation results.

TABLE 2.1   Effectiveness of Sub-Basin Division in Class I WRRs

Class I WRRs	Average Area/km²	Coincidence Value/%	Class I WRRs	Average Area/km²	Coincidence Value/%
SRB	305.18	99.82	YZRB	431.40	100.13
LRB	271.63	99.79	SERB	160.92	99.73
HRB	438.27	99.92	PRB	274.82	99.81
YRB	434.46	100.08	SWRB	389.14	99.86
HURB	307.85	100.02	NWRB	409.01	99.89

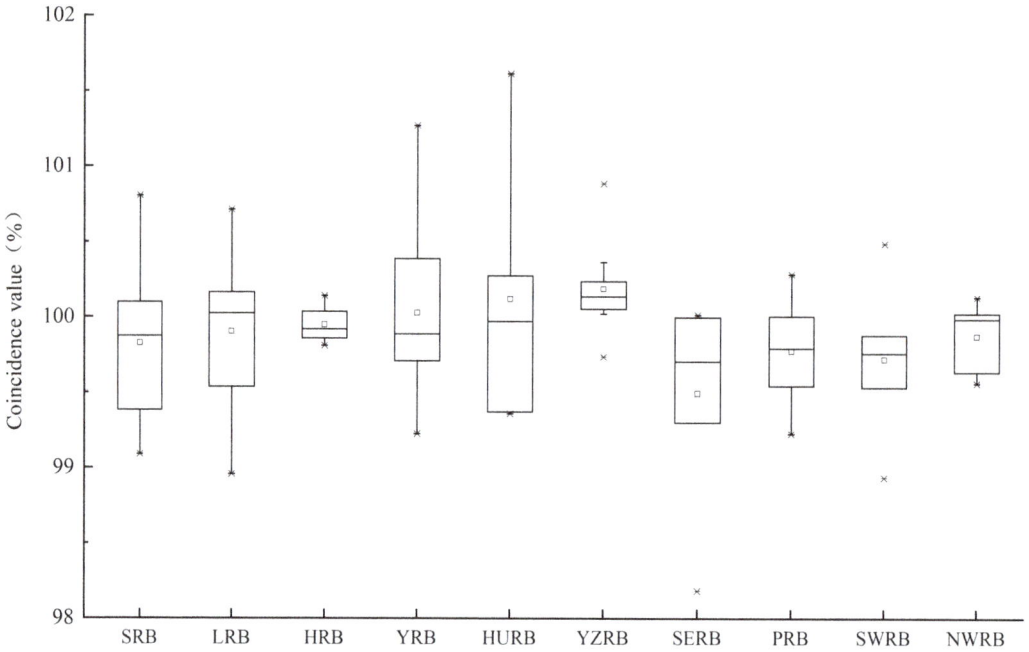

FIGURE 2.15 Box plot of the change in the extent to which the sub-basins' coverage of 80 Class II WRRs coincides with its actual extent.

REFERENCES

Barnes, R., Lehman, C. and Mulla, D., 2014. An efficient assignment of drainage direction over flat surfaces in raster digital elevation models. Computers and geosciences, 62, 128–135.

Jenson, S.K. and Domingue, J.O., 1988. Extracting topographic structure from digital elevation data for geographic information system analysis. Photogrammetric engineering and remote sensing, 54(11), 1593–1600.

Liu, J., Zhou, Z., Jia, Y., et al., 2014. A stem-branch topological codification for watershed subdivision and identification to support distributed hydrological modeling at large river basins. Hydrological processes, 28(4), 2074–2081.

Reddy, G.P.O., Kumar, N., Sahu, N., et al., 2018. Evaluation of automatic drainage extraction thresholds using ASTER GDEM and Cartosat-1 DEM: A case study from basaltic terrain of Central India. The Egyptian journal of remote sensing and space science, 21(1), 95–104.

Sousa, T.M.I. and Paz, A.R., 2017. How to evaluate the quality of coarse-resolution DEM-derived drainage networks. Hydrological processes, 31, 3379–3395.

CHAPTER 3

Infiltration-Runoff Modeling for Swelling Soil under Unsteady Rainfall Condition

3.1 RESEARCH BACKGROUND

Rainfall-infiltration is the dynamic process of water moving downward through the surface to supply soil water and groundwater, which is an important component of hydrological processes. Due to its importance in the hydrological cycle, infiltration has become one of the most intensively investigated topics in hydrology and soil physics. Currently, focusing on the study of water movement in non-swelling soils, a large number of laboratory and field experiments have been carried out by domestic and foreign scholars, and many mathematical models have been established (Green and Ampt, 1911; Gan et al., 2015; Vatankhah, 2015). However, swelling soils swell when absorbing water and shrink when desorbing water. Volume change associated with water movement is central to these issues, but present approaches to the hydrology of swelling soils were generally based on non-swelling soil theory, and volume change of soil and its consequences were rarely considered. Consequently, significant errors may arise in estimations of local water and solute flux, water balance, and aquifer recharge.

Swelling soils, the special soil type distributed widely all around the world, exhibit distinct swelling and shrinking properties, causing the uplift or subsidence of structure foundations, and resulting in engineering disasters and great economic losses (Azam et al., 2013). The research on water movement in swelling soils has become the focus of attention for some disciplines, such as engineering geology, hydrology, and agrology. Since the one-dimensional consolidation theory of Terzaghi (1925), there have been many attempts to model soil deformation and water flow. Based on the measurement of water and solid particle movement. Kim et al. (1992) proposed a numerical model accounting for one-dimensional soil deformation and associated water flow based on the governing equation developed by Philip (1968). For two-dimensional analysis of soil deformation, a geometry

DOI: 10.1201/9781003646648-3

factor (Bronswijk, 1990) was introduced to obtain vertical and horizontal components of shrinkage and swelling from the SSC of a soil. Based on the geometry factor, Kim et al. (1999) showed that the model could be applied to the two-dimensional soil deformation system in a shrinking condition. The model could be applied to simulate soil deformation and water flow under saturated/unsaturated conditions and internal/external load application with two-dimensional analysis. All these models presumed that Darcy's law, which focuses on one-dimensional flow, is valid for swelling soils (Philip, 1968). However, the solution of these equations requires an iterative implicit numerical technique with fine discretization in time and space, which was computationally intensive (Rao et al., 2006). So, it is time-consuming when fully distributed simulations are applied at a large scale. In addition, some of its basic concepts remain misunderstood.

To explain the mechanics widely known as anomalous diffusion, Su (2014) and Lockington and Parlange (2003) used a fractional partial differential equation (fFPE) to analyze swelling soil water flow. It provided a fresh explanation and understanding of infiltration into swelling media. The theory of water flow in rigid soils has been used for more than 60 years. The water flow in swelling soil is not well established in soil science, although it is increasingly considered in engineering geology, hydrology, and agrology. The swelling soil deformation when absorbing water is mainly affected by the soil swelling pressure and self-weight stress. The soil swelling pressure varies with soil water content, and the self-weight stress varies with soil depth. The soil deformation characteristics under pressure change with the increase of soil depth, which causes the change of saturated water movement parameters, including the soil saturated hydraulic conductivity and soil saturated water content. The parameters strongly influence the infiltration process of soil water. However, current studies have not well established the theory of soil water movement parameters and infiltration processes in swelling soils.

To investigate the effects of soil swelling deformation, the soil saturated hydraulic conductivity and soil saturated water content considering soil swelling deformation were introduced, and an infiltration-runoff model considering swelling deformation effects under unsteady rainfall (GJGAM) was established based on the Green-Ampt model in this study. Then, the Loessial soil and Lou soil were selected as typical swelling soils, and one-dimensional infiltration-runoff experiments under a series of different soil thicknesses were performed to verify the GJGAM. The research takes a fresh explanation and understanding of the infiltration-runoff process into swelling media and provides some guidance in managing and regulating swelling soil water.

3.2 MATERIAL AND METHODS

3.2.1 Laboratory Experiment for Soil Swelling Deformation Effects on Saturated Water Movement Parameters

3.2.1.1 Testing Soil Samples

The saturated water movement parameters that consider soil swelling deformation effects, including the soil saturated hydraulic conductivity and soil saturated water content, were introduced to establish the GJGAM.

TABLE 3.1   Basic Physical Properties of Soil

| Soil Type | Particle-size Distribution/% | | | ρ_s/(g/cm³) | ρ_d/(g/cm³) | K_0/(cm/min) |
	>50μm	2–50μm	<2μm			
Loessial Soil	3.50	52.00	44.50	1.40	2.628	0.014
Lou Soil	31.30	46.20	22.50	1.40	2.669	0.017

Considering the swelling properties of loess soils, the Loessial soil and Lou soil were selected as the typical swelling soils. Each of the soils was air-dried and then sieved through a 2 mm screen. The air-dried and sieved soils were prepared for testing soil samples. The soil particle-size distribution was measured using a laser particle size analyzer (Table 3.1). As two typical swelling soils, the properties of Loessial soil and Lou soil are listed in Table 3.1. The experimental water was running water supplied by the Institute of Soil and Water Conservation, CAS & MWR.

3.2.1.2 Testing Device

The schematic representation of the experimental setup is shown in Figure 3.1. It included a packed column, Marriott tube, and weighing equipment. The experiments were performed on a uniformly packed column. The inner diameter of the columns was 10 cm, the length of the columns was 4, 6, 8, 10, 15, 20, 25, 30, 35, 40, 45, 50, 55, and 60 cm, respectively.

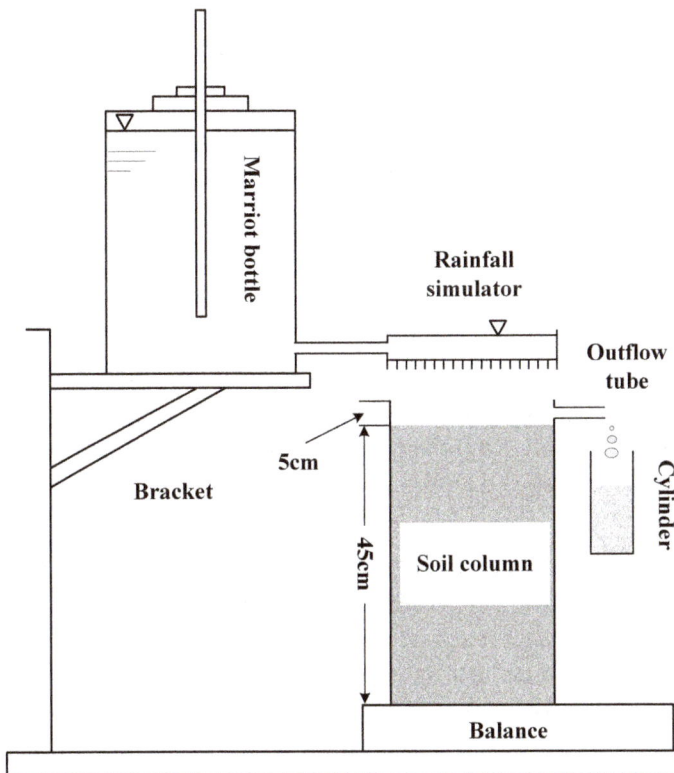

FIGURE 3.1   Schematic diagram of the experimental setup.

A Marriott tube was used to supply water (inner diameter 10 cm, height 50 cm). A 2.5 cm fine sand layer was located at the bottom of the column, in which a perforated PVC pipe was embedded to allow for the water supply.

3.2.1.3 Testing Design

A series of different soil thicknesses was set to determine the saturated water movement parameters at each soil thickness, and then the parameters at different soil depths were calculated approximately. The soil saturated water content was calculated by the mass conservation law, while the soil saturated hydraulic conductivity was calculated by the flux equivalence principle (harmonic mean). Using Loessial soil and Lou soil as the testing soils, the laboratory experiments were designed with three replications of 14 thicknesses of 2, 4, 6, 8, 10, 15, 20, 25, 30, 35, 40, 45, 50, and 55 cm, respectively.

3.2.1.4 Testing Procedure

(1) Soil Packing   After being sieved through a 2 mm screen, the air-dried soils were compacted into a uniformly packed column with a targeted bulk density (1.4g/cm³) in incremental stages (2 cm for the total thickness less than 10 cm and 5 cm for the total thickness larger than 5 cm). The soil thicknesses (i.e., the height of the soil column) were 2, 4, 6, 8, 10, 15, 20, 25, 30, 35, 40, 45, 50, and 55 cm, respectively.

(2) Outflow Equipment Installation   After the soils were packed uniformly, the bottom of the soil column was welded with a PVC plate to ensure that the welded joint was not leaking. The PVC plate was composed of an upper and lower plate, which were connected with a PVC ring (inner diameter 10 cm, wall thickness 0.5 cm). The upper plate was covered with a circular hole with a diameter of about 1 mm to allow for the water supply and soil support. The upper surface of the lower plate was concave, and the lower surface was plane. The center position of the lower plate was embedded with a perforated PVC pipe (inner diameter 1 cm) to facilitate the water collection.

(3) Water Supply Equipment Installation   After the outflow equipment was installed, the Marriott tube with a graduated scale (inner diameter 10 cm, height 50 cm) was used to supply water through a PVC pipe from the outflow of the PVC plate. The bubbling point head and the soil column surface were on the same horizontal plane.

(4) Experimental Data Observation   At the point when the soil sample was fully expanded and saturated, the experiment was started. All of the tested samples were prepared using identical procedures relative to packing. The soil saturated conductivity was determined using a constant-head method (when the thickness was less than 10 cm, the water ponding depth was 4 cm, and when the thickness was larger than 10 cm, the water ponding depth was 10 cm). The soil saturated water content was determined using a soak test (Klute, 1986). The soil swelling deformation was measured using a Vernier caliper.

3.2.2 Laboratory Experiment for Soil Swelling Deformation Effects on Infiltration-Runoff

3.2.2.1 Testing Soil Samples

Like the laboratory experiment for soil swelling deformation effects on saturated water movement parameters, Loessial soil and Lou soil were selected as the typical swelling soils. The soils were air-dried and then sieved through a 5 mm screen, respectively. The air-dried and sieved soils were prepared for testing soil samples. The experimental water was running water supplied by the Institute of Soil and Water Conservation, CAS & MWR.

3.2.2.2 Testing Design

The experiments were performed in an experimental steel box. The box is 0.7m high, 1m wide, 1m long, and 5° slope steepness. The box has two water outlets, upper and lower water outlets, respectively. The surface runoff outflowed through the upper outlet, and the subsurface runoff outflowed through the lower outlet.

The soil swelling deformation on infiltration-runoff experiments was designed with two replications of four thicknesses of 10, 20, 30, and 40 cm, respectively. The Loessial soil with four thicknesses of 10, 20, 30, and 40 cm was named as 10 cm Loessial soil, 20 cm Loessial soil, 30 cm Loessial soil, and 40 cm Loessial soil, respectively. Similarly, the Lou soil with four thicknesses of 10, 20, 30, and 40 cm was named as 10 cm Lou soil, 20 cm Lou soil, 30 cm Lou soil, and 40 cm Lou soil, respectively.

3.2.2.3 Testing Procedure

Before filling the box with soil, a 10 cm fine sand layer taking filter action was packed at the bottom of the box, and then the gauze was laid above the fine sand. The air-dried and sieved through a 5 mm screen soil sample was compacted into the experimental steel box in incremental stages (10 cm) with a targeted bulk density (1.4 g/cm^3). During the compacting process, four Trime soil moisture sensors were installed in the middle position of the steel box to measure the soil water dynamic process at the depths of 0, 10, 25, and 40 cm.

Following the soil sample being set up in the box, enough water was applied until the soil water content was larger than the field water capacity (the bottom steel pipe outflowed continuously). The soil sample was then covered with plastic to prevent evaporation and left undisturbed until fully deformed.

At the point when the soil samples were fully deformed, the infiltration-runoff experiments began. During the experiments, the rainfall system was used to carry out artificial rainfall, and the soil water dynamics were measured and recorded by the Trime system. The rainfall rate was 30 mm/h, and the duration of the rainfall was approximately 350 minutes. The observed variables were rainfall rate (mm), surface flow rate (m^3/s), subsurface flow rate (m^3/s), and rainfall duration (min). When the rainfall stopped, the soil's bulk density was measured using an oven-dry method (100 cm^3 cutting ring).

To conveniently analyze the infiltration process, the soil water characteristic curve was measured using the GR21G centrifuge and fitted with the Van Genuchten model (Van Genuchten, 1980) (Table 3.2). The soil swelling characteristic curve was measured using a Vernier caliper and fitted with the three straight lines model (Table 3.3). As the soil used in

TABLE 3.2   The Fitted Parameters of the Van Genuchten Model

Soil Type	α	m	n	R^2
Loessial Soil	0.020	0.266	1.363	0.99
Lou Soil	0.014	0.152	1.179	0.99

TABLE 3.3   The Fitted Parameters of the Three Straight Lines Model

Soil Type	a	α_1	R^2	b	α_2	R^2	c	α_3	R^2	U_A	U_B	U_S
Loessial Soil	0.70	0.05	0.96	0.66	0.24	1.00	0.73	0.03	1.00	0.23	0.32	0.34
Lou Soil	0.6	0.25	0.99	0.57	0.47	0.98	0.59	0.35	0.98	0.18	0.23	0.31

this experiment was the same as the testing soil used by Huang and Shao (2008), the soil swelling characteristic curve was obtained directly from their results. The soil stress-strain relationship curve was measured by the constant pressure method and fitted with the logarithmic function (Table 3.4). Since the depth of the soil column was less affected by rainfall (less than 1m), the self-weight stress was less than 25kPa, so the constant pressure range was 0–25kPa. The soil bulk density was determined by the density bottle method. The bulk density and the corresponding soil saturated hydraulic conductivity (bulk density 1.1, 1.2, 1.3, 1.4, and 1.5 g/cm³) were determined by the constant volume method and fitted with the model proposed by Lambe and Whiman (1979) (Table 3.5).

The relationship between soil water content and soil water suction was described using the van Genuchten model, and the model could be calculated by equation (3.1).

$$S = \frac{\theta - \theta_r}{\theta_s - \theta_r} = \left[\frac{1}{1 + (\alpha h)^n} \right]^m \tag{3.1}$$

where h is the soil suction, cm; θ is the soil water content, cm³/cm³; θ_s is the saturated water content, cm³/cm³; θ_r is the residual water content, cm³/cm³; and α, m, and n represent the parameters, $m = 1 - 1/n$.

The soil swelling characteristic curve was described using the three straight lines model (McGarry and Malafant, 1987) and could be given as equation (3.2).

$$\begin{cases} v = a + \alpha_1 U & 0 < U < U_A \\ v = b + \alpha_2 U & U_A < U < U_B \\ v = c + \alpha_3 U & U_B < U < U_S \end{cases} \tag{3.2}$$

TABLE 3.4   Soil Stress-Strain Relationship Curve Parameters

Soil Type	A	B	R^2
Loessial Soil	1.093	0.1044	0.99
Lou Soil	1.023	0.1216	0.99

TABLE 3.5   The Fitted Parameters of the Model Proposed by Lambe and Whiman (1979)

Soil Type	Saturated Hydraulic Conductivity/(cm/min)	m	R^2
Loessial Soil	0.019	6.18	0.97
Lou Soil	0.017	12.2	0.98

where ν is the specific volume that is the reciprocal of the soil bulk density, cm^3/g; U is the mass water content, g/g; α_1, α_2, and α_3 are the slope of soil swelling characteristic curve; U_A, U_B, and U_S are the mass water content at the inflection point of the curve, g/g; a, b, and c represent the parameters.

The soil stress-strain relationship curve could be described using a logarithmic equation (3.3).

$$\rho_s = A + B \ln p \tag{3.3}$$

where ρ_s is the soil bulk density, g/cm^3; p is the stress, N/cm^2; A and B represent the parameters.

The relationship between the soil saturated hydraulic conductivity and soil porosity is determined by Lambe and Whiman (1979):

$$K_s(e) = K_0 10^{m(e-e_0)} \tag{3.4}$$

where $K_S(e)$ is the soil saturated hydraulic conductivity when soil porosity is e, cm/min; K_0 is the soil saturated hydraulic conductivity when soil porosity is e_0, cm/min; m is the parameter related to the properties of soil porosity.

3.3 MODELING PROCESS IMPROVEMENTS

3.3.1 Swelling Soil Saturated Water Movement Parameters Calculating Models

3.3.1.1 Description of Soil Deformation Process

(1) The Influence of Soil Swelling Pressure on Soil Deformation   Without consideration of the self-weight stress, when absorbing water, the soil will deform freely under the action of the soil swelling pressure. The swelling deformation is a function of the soil water content, and the deformation could be calculated by the three straight line model (see equation (3.2)).

(2) The Influence of Soil Self-Weight Stress on Soil Deformation   Under the action of soil self-weight stress, the soil deformation process could be described using the logarithmic function, and equation (3.3) can be modified as:

$$\rho_s = A + B \ln \gamma z \tag{3.5}$$

$$\gamma = \rho_w g \tag{3.6}$$

where ρ_s is the soil bulk density, g/cm^3; γ is the wet bulk density, N/cm^3; z is the coordinate axis; ρ_w is the wet bulk density, g/cm^3; g is the gravitational acceleration, g/N; A and B represent the parameters.

Under the joint force of the soil swelling pressure and self-weight stress, the soil porosity caused by the swelling deformation of the soil varies with soil depth. Assuming that the volume occupied by soil is constant in the swelling deformation process of unit soil, the soil deformation is equivalent to the change of soil porosity. Furthermore, the change of soil porosity could be expressed as equations (3.7) and (3.8).

$$\begin{cases} de = de_w + de_p \\ de_w = \mu dU \\ de_p = \gamma \beta dz \end{cases} \tag{3.7}$$

$$\mu = \frac{de}{dU}, \quad \beta = \frac{de}{d\sigma} = \frac{de}{\gamma dz} \tag{3.8}$$

where e is the soil porosity, cm³/cm³; e_w is the soil porosity caused by the soil swelling deformation, cm³/cm³; e_p is the soil porosity caused by the soil self-weight stress, cm³/cm³; U is the mass water content, g/g; μ is the slope of soil swelling characteristic curve; β is the slope of soil stress-strain relationship curve; σ is the soil self-weight stress, N/cm²; γ is the wet bulk density, N/cm³; z is the soil depth, cm.

3.3.1.2 Soil Saturated Water Content Calculating Model

The soil swelling deformation is caused by the change in soil porosity. Consequently, when the soil is saturated, according to equation (3.7), the change of soil porosity due to the soil swelling pressure could be expressed as equation (3.9).

$$\Delta e_w = 1 - \frac{\rho_s}{\rho_d} - e_0 = (1 - e_0) - \frac{1}{\rho_d(c + \alpha U)} \tag{3.9}$$

Similarly, according to equation (3.7), the change of soil porosity due to the soil self-weight stress could be expressed as equation (3.10).

$$\Delta e_p = 1 - \frac{\rho_s}{\rho_d} - e_0 = (1 - e_0) - \frac{A + B\ln(\gamma z)}{\rho_d} \tag{3.10}$$

where ρ_d is the soil particle density, g/cm³; e_0 is the soil initial porosity, cm³/cm³; the other symbols have the same meaning as described above.

As the soil is saturated, the soil pores are filled with water. The soil saturated water content is equivalent to the soil porosity, so the soil saturated water content in the whole soil profile could be calculated as equation (3.11).

$$\theta_T = \int_0^z e\, dz = \int_0^z (e_0 + \Delta e_w + \Delta e_p)\, dz$$

$$= (2 - e_0)z - \frac{(A - B)z}{\rho_d} - \frac{Bz\ln(\gamma z)}{\rho_d} - \frac{z}{\rho_d(c + \alpha_3 U_S)} \tag{3.11}$$

where θ_T is the saturated water content of the region above soil depth z, cm³/cm³; z is the soil depth, cm; c and α_3 represent the fitted parameters in the saturated phase of the three straight lines model; the other symbols have the same meaning as described above.

3.3.1.3 Soil Saturated Hydraulic Conductivity Calculating Model

Considering the relationship between the soil saturated hydraulic conductivity and soil depth, the model could be modified as equation (3.12).

$$K_{sz}(e) = K_0 10^{m(e_z - e_0)} \tag{3.12}$$

$$e_z = 2 - e_0 - \frac{1}{\rho_d(c + \alpha_3 U)} - \frac{A + B\ln(\gamma z)}{\rho_d} \tag{3.13}$$

where $K_{sz}(e)$ is the soil saturated hydraulic conductivity when soil porosity is e_z, cm/min; e_z is the soil porosity at soil depth z, cm³/cm³; e_0 is the soil initial porosity, cm³/cm³; K_0 is the soil saturated hydraulic conductivity when soil porosity is e_0, cm/min; m is the parameter related to the properties of soil porosity.

3.3.2 Swelling Soil Infiltration-Runoff Model

The schematic diagram of the infiltration process into the soil profile is illustrated in Figure 3.2. The origin of the coordinate system was set at the soil surface, and the coordinate system was positively downward. Soil deformation occurs when swelling soils absorb water. Both the soil saturated water content and soil saturated hydraulic conductivity in the wetting zone change with soil depth. To facilitate the analysis, some assumptions need to be made: the swelling soil is homogeneous before deformation; the soil swelling deformation is elastic, which means that there is no hysteresis in the deformation; the soil swelling deformation only caused the change of soil porosity. There is a specific wetting front in the infiltration process, and the front separates the wetting zone and non-wetting zone. The soil is saturated in the wetting zone above the wetting front, while the soil water

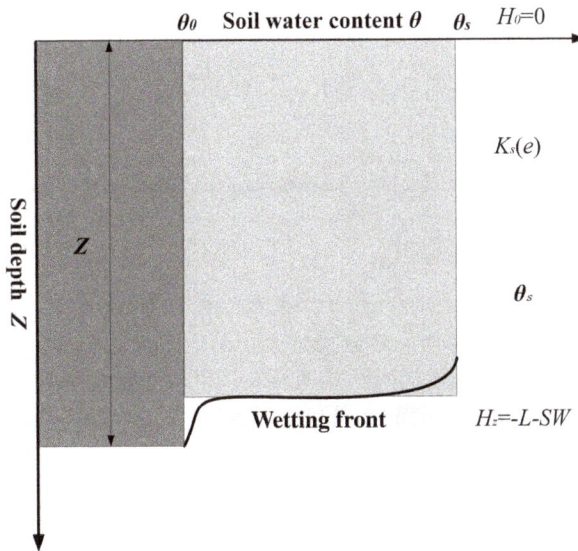

FIGURE 3.2   Schematic diagram of the infiltration process in the soil profile.

content in the non-wetting zone below the wetting front is soil initial water content. In order to describe the change characteristics of soil hydraulic conductivity with soil depth, the saturated hydraulic conductivity for swelling soil $K_{sz}(e)$ was introduced. Moreover, the saturated water content for swelling soil (θ_T) was introduced to describe the soil saturated water content in the wetting zone. Consequently, an infiltration-runoff model for swelling soils, the GJGAM (a modified Green-Ampt model), was established based on the Green-Ampt model. Compared with the TGAM, the GJGAM considered the change of the soil saturated water content and soil saturated hydraulic conductivity caused by soil swelling deformation.

For the infiltration under unsteady rainfall conditions, the rainfall process was divided into n periods of equal length, within which the rainfall rate was constant. The infiltration process at time $(t_{x-1} \sim t_x)$ was determined by ponding depth h_0 at the start time of the period, rainfall rate I, and ponding infiltration rate f_{pt}, where t_x is the time at x, and n represents the number of the discrete periods occurring in the rainfall record. Since the unit of ponding depth h_0 is different from that of the rainfall intensity, it is necessary to first divide the ponding depth h_0 by the corresponding time interval Δt during the calculation process, and convert it to the rainfall intensity P'.

As shown in Figure 3.3, according to the ponding depth h_0 at the start time of the period, the rainfall intensity P', and the ponding infiltration rate f_{pt}, the infiltration process during the period could be divided into the following scenarios:

Case a: $h_0 = 0$, $I > f_{pt} \geq K_{sz}(e)$. In this situation, surface ponding will occur with the sustained rainfall, and the infiltration process can be divided into non-ponding infiltration f_{npt} (before t_p), and ponding infiltration f_{pt} (after t_p).

Case b: $h_0 > 0$, $P' + I < K_{sz}(e) \leq f_{pt}$. In this situation, no ponding will occur with the infiltration process, and the process can be divided into ponding infiltration f_{pt} (before t_p), and non-ponding infiltration f_{npt} (after t_p).

Case c: $h_0 > 0$, $P' + I \geq f_{pt} \geq K_{sz}(e)$. In this situation, the infiltration process is under a ponding condition f_{pt}.

Case d: $h_0 = 0$, $I < K_{sz}(e) \leq f_{pt}$. In this situation, the infiltration process is under a non-ponding condition f_{npt}.

where $K_{sz}(e)$ is the swelling soil hydraulic conductivity in the wetting zone; I is the rainfall intensity; h_0 is the ponding depth at the start time of the period; f_{npt} is the non-ponding infiltration rate; f_{pt} is the ponding infiltration rate; t_p is the time surface ponding occurs.

According to Darcy's law, it could be obtained using equations (3.14)–(3.16).

Before the ponding occurs:

$$\begin{cases} f_{npt} = I \\ F = F_{x-1} + (t - t_{x-1})I \end{cases} \tag{3.14}$$

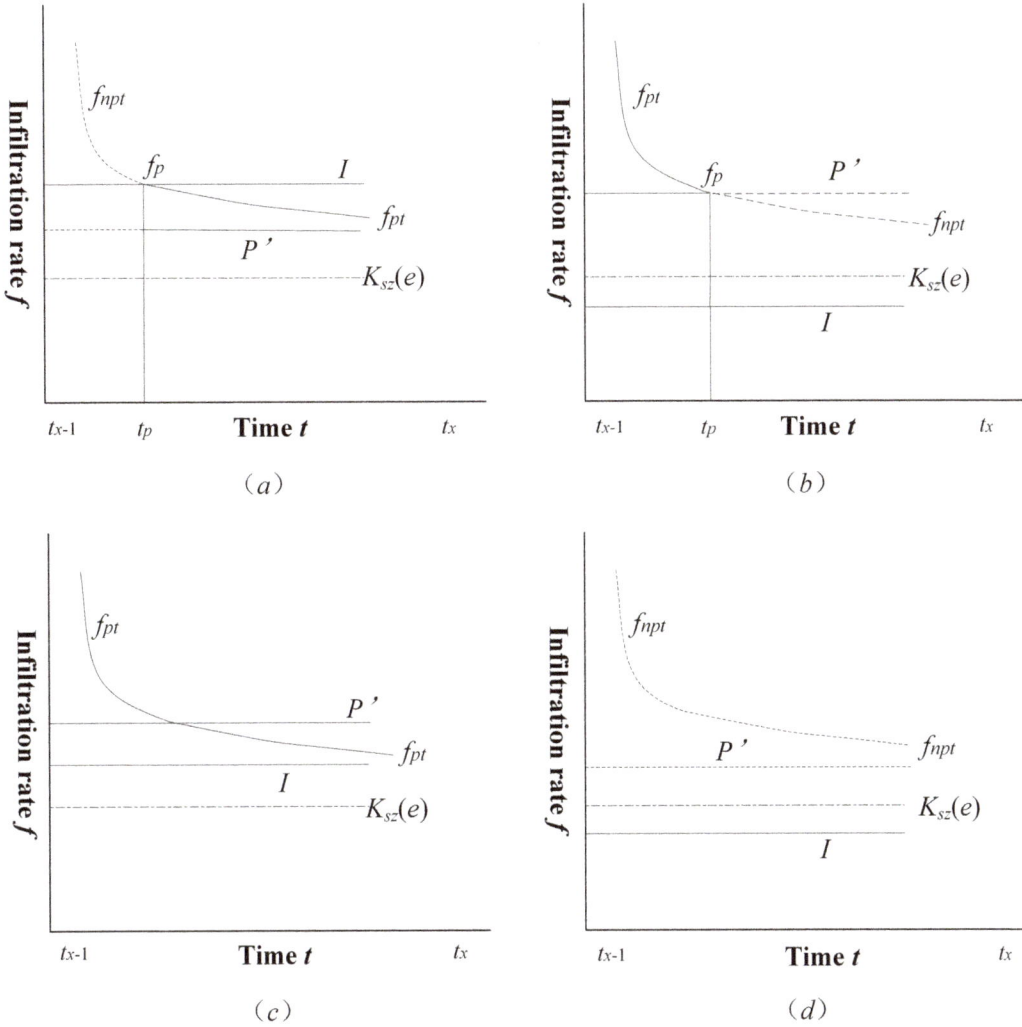

FIGURE 3.3 Shift between the ponding and non-ponding conditions: *a*, *b*, *c*, and *d* represent the different infiltration processes under the ponding and non-ponding conditions.

After the ponding occurs:

$$f_p = -K_{sz}(e)\frac{-(SW+Z)-H_0}{Z} \tag{3.15}$$

Neglecting the surface ponding:

$$f_p = -K_{sz}(e)\frac{-(SW+Z)}{Z} \tag{3.16}$$

where f_{npt} is the infiltration rate before the ponding occurs, cm/min; I is the rainfall intensity, cm/min; f_{pt} is the infiltration rate after the ponding occurs, cm/min; SW is the soil suction head at wetting front, cm; H_0 is the surface ponding depth, cm; Z is the wetting front distance, cm; the other symbols have the same meaning as described above.

Next, according to the mass balance equation, the cumulative infiltration F at time t was calculated as equation (3.17).

$$F = \theta_T - \theta_0 Z = \left[(2 - e_0) - \frac{A - B}{\rho_d} - \frac{B\ln(\gamma z)}{\rho_d} - \frac{1}{\rho_d(r + sU)} - \theta_0 \right] Z \tag{3.17}$$

Let:

$$\Delta\theta = (2 - e_0) - \frac{A - B}{\rho_d} - \frac{B\ln(\gamma z)}{\rho_d} - \frac{1}{\rho_d(r + sU)} - \theta_0 \tag{3.18}$$

Then:

$$Z = \frac{F}{\Delta\theta} \tag{3.19}$$

where θ_T is the saturated water content of the region above the wetting front, cm³/cm³; F is the cumulative infiltration, cm.

$$f_{pt} = \frac{dF}{dt} = K_{sz}(e)\left[1 + \frac{SW\Delta\theta}{F} \right] \tag{3.20}$$

The following could be obtained by an integral solution:

$$F - F_p = K_{sz}(e)(t - t_p) + A \cdot \ln\left[\frac{A + F}{A + F_p} \right] \tag{3.21}$$

$$A = SW\Delta\theta \tag{3.22}$$

The time surface ponding occurs t_p was defined as follows:

$$t_p = t_{x-1} + \frac{F_p - F_{n-1}}{I_p} \tag{3.23}$$

$$F_p = \frac{A}{I_p \Big/ K_{sz}(e) - 1} \tag{3.24}$$

where F_p is the accumulative infiltration at t_p, cm; t_p is the time surface ponding occurs, min; I_p is the rainfall intensity during the xth time step within which surface ponding occurs, cm/min; θ_0 is the soil initial water content, cm³/cm³; SW is the average soil suction head at wetting front, cm; A represents the parameter; the other symbols have the same meaning as described above.

3.3.3 Model Parameters

The model parameters of GJGAM consisted of the soil saturated water content, soil initial water content, soil saturated hydraulic conductivity, and soil suction head at the wetting front. To calculate saturated water content, the soil swelling characteristic curve, soil stress-strain relationship curve, and wet bulk density were determined. When calculating the soil saturated hydraulic conductivity of swelling soils, except for the soil swelling characteristic curve, soil stress-strain relationship curve, and wet bulk density (γ) that are to be determined, it is also necessary to determine the relationship between soil porosity and the hydraulic conductivity. The soil initial water content at different soil depths was measured, and the soil suction head (SW) at the wetting front was determined by the van Genuchten model and half of the air-bubbling capillary pressure (Bouwer, 1969). If the soil is saturated, the soil suction head (SW) at the wetting front is 0. In addition, the wet bulk density (γ) was replaced by the soil saturated bulk density that was measured by the constant volume method.

In the TGAM, the soil structure was considered to remain constant during the wet-dry alternation, and both the soil saturated water content and soil saturated hydraulic conductivity were unchanged with soil depth. Consequently, the soil saturated water content and soil saturated hydraulic conductivity corresponding to the soil initial bulk density were used in the TGAM and determined by the constant volume method. Moreover, the soil initial water content at different soil depths and the soil suction head (SW) at the wetting front were determined, and the methods were the same as those of the GJGAM.

The parameters of the calculated model of soil saturated water content and soil saturated hydraulic conductivity considering soil swelling deformation, and the input parameters of the GJGAM for the laboratory experiments are summarized in Table 3.6. The model input parameters of the TGAM are listed in Table 3.7.

Based on the previously determined parameters, the calculated soil saturated water content and soil saturated hydraulic conductivity considering soil swelling deformation were tested by the observed data. In addition, the cumulative infiltration of swelling soils was simulated by the GJGAM and TGAM, and then compared with the observed data. To compare the observed and measured values, the average relative error (ARE), Nash-Sutcliffe

TABLE 3.6   The Model Input Parameters of the GJGAM

Soil Type	c	α_3	U_s	A	B	m	ρ_d/(g/cm³)	ρ_w/(g/cm³)	SW/cm	K_0/(cm/min)
Loessial Soil	0.73	0.03	0.34	1.093	0.104	6.18	2.669	1.87	20	0.014
Lou Soil	0.59	0.35	0.31	1.023	0.121	12.2	2.628	1.87	35	0.017

TABLE 3.7   The Model Input Parameters of the TGAM

Soil Type	Soil Saturated Water Content/(cm^3/cm^3)	Soil Saturated Hydraulic Conductivity/(cm/min)	Soil Suction/cm
Loessial Soil	0.47	0.014	20
Lou Soil	0.47	0.017	35

efficiency coefficient (*NSE*), and root mean square error (*RMSE*) were used as the three criteria to reflect the simulation effectiveness of the models.

$$ARE = \frac{1}{n}\sum_{1}^{n}\left(\frac{Q_{p,i}-Q_{o,i}}{Q_{o,i}}\right) \times 100 \tag{3.25}$$

$$NSE = 1 - \frac{\sum_{1}^{n}(Q_{o,i}-Q_{p,i})^2}{\sum_{1}^{n}(Q_{o,i}-\overline{Q_o})^2} \tag{3.26}$$

$$RMSE = \sqrt{\frac{\sum_{1}^{n}(Q_{o,i}-Q_{p,i})^2}{n}} \tag{3.27}$$

where n is the number of observations; $Q_{o,i}$ and $Q_{p,i}$ are the observed and predicted values, respectively; $\overline{Q_o}$ represents the mean values of $Q_{o,i}$.

3.4 RESULTS AND DISCUSSION

3.4.1 Swelling Soil Saturated Water Movement Parameters

The *ARE, RSME,* and *NSE* values between the simulated data and observed data of swelling soil saturated water movement parameters, the soil saturated water content, and soil saturated hydraulic conductivity, are shown in Table 3.8. The *ARE* values were less than 10%, the *RSME* values were less than 0.07, and the *NSE* values were larger than 0.85, whether for Loessial soil or Lou soil. The goodness of fit statistics indicates a reasonable agreement between the observed and simulated parameters.

Changes of measured and simulated values of the saturated water content of Loessial soil and Lou soil with soil thickness are shown in Figure 3.4. It can be seen that the soil saturated water content decreases with increasing soil thickness. Because Loessial soil and Lou soil are swelling soils, their swelling deformation when absorbing water is mainly affected by the soil swelling pressure and self-weight stress. With soil depth increases, the self-weight stress increases and plays an increasing role, causing the soil bulk density increases. And then the soil porosity and soil saturated water content decrease with increasing soil bulk density. Consequently, with the soil thickness increases, the soil saturated water content decreases.

TABLE 3.8   The *ARE, RSME,* and *NSE* Values of the Simulated Data and Observed Data of Swelling Soil Saturated Water Movement Parameters

Soil Type	Soil Saturated Movement Parameters	*ARE*/(%)	*RSME*/(cm/min)	*NSE*
Loessial Soil	Soil saturated hydraulic conductivity /(cm/min)	−7.61	0.96	0.001
	Soil saturated water content/(cm³/cm³)	−0.57	0.95	0.006
Lou Soil	Soil saturated hydraulic conductivity /(cm/min)	3.05	0.94	0.001
	Soil saturated water content /(cm/min)	3.21	0.93	0.030

(a) Loessial soil

(b) Loessial soil

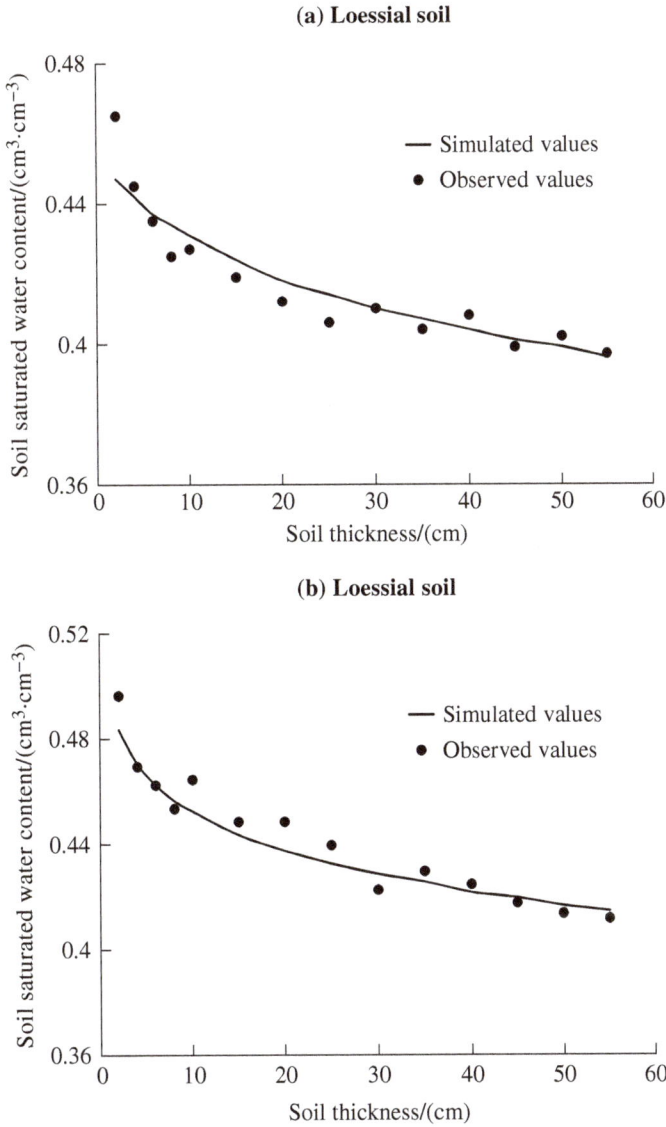

FIGURE 3.4 Changes of measured and simulated values of the saturated water content of Loessial soil and Lou soil with soil thickness.

Changes of measured and simulated values of the saturated hydraulic conductivity of Loessial soil and Lou soil with soil thickness are shown in Figure 3.5. Because the soil porosity increases with increasing soil thickness, the soil saturated hydraulic conductivity decreases gradually.

3.4.2 Soil Cumulative Infiltration

The *ARE*, *RSME*, and *NSE* values of the GJGAM and TGAM in simulating soil cumulative infiltration are shown in Table 3.9. For all soil depth treatments, the *ARE* values

(a) Loessial soil

(b) Lou soil

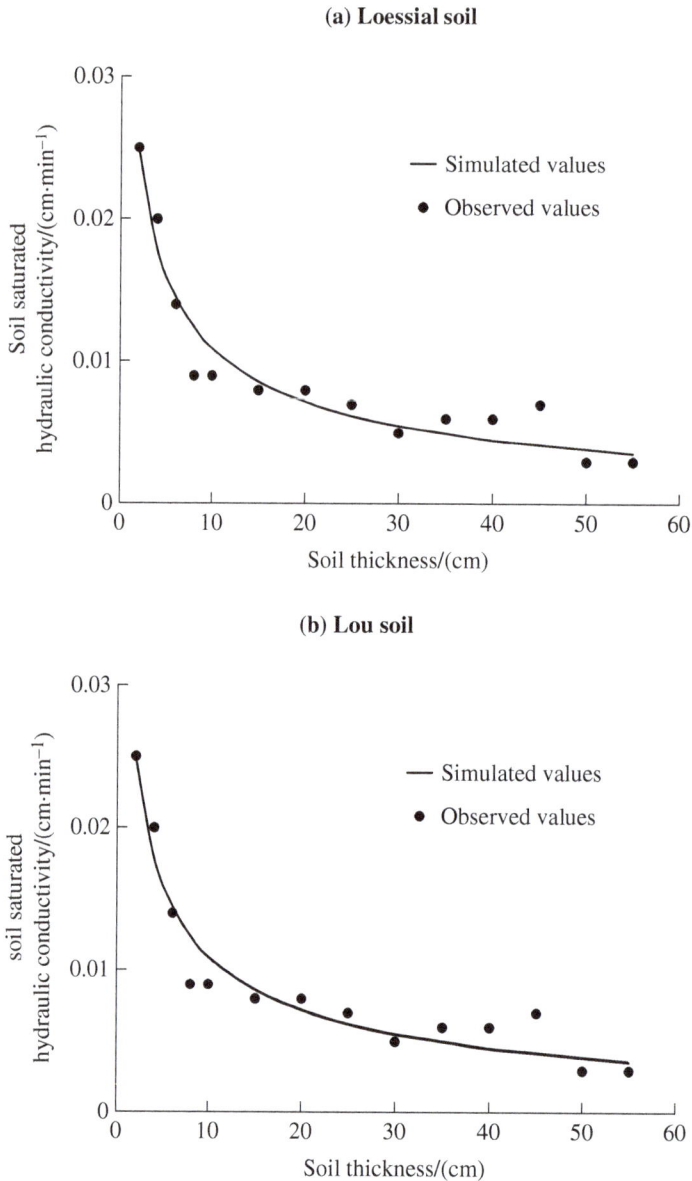

FIGURE 3.5   Changes of measured and simulated values of the saturated hydraulic conductivity of loessial soil and lou soil with soil thickness.

of the GJGAM were –7.99% to 18.86%, the *RSME* values were less than 0.42 cm, and the *NSE* values were larger than 0.79. However, when using the TGAM, the *ARE* values were –6.70% to 139.72%, the *RSME* values were larger than 0.20 cm, and the *NSE* values were –17.55 to 0.99. In comparison, the GJGAM was more suitable for simulating the infiltration process of swelling soils. TGAM failed to consider the effect of soil swelling deformation on the infiltration process, which caused the cumulative infiltration to be overestimated.

TABLE 3.9 The *ARE*, *RSME*, and *NSE* Values of the Soil Cumulative Infiltration of Soil between the Simulated Data of TGAM and GJGAM and the Observed Results

Model	Soil Depth Treatment	Lou Soil			Loessial Soil		
		ARE/(%)	*RSME*/(cm/min)	*NSE*	*ARE*/(%)	*RSME*/(cm/min)	*NSE*
TGAM	10 cm	17.95	0.35	0.96	−6.7	0.62	0.93
	20 cm	53.65	1.56	−4.11	4.45	0.2	0.99
	30 cm	139.72	2.2	−17.55	12.91	0.63	0.91
	40 cm	97.77	1.84	−11.84	23.15	0.76	0.73
GJGAM	10 cm	−7.99	0.22	0.98	0.59	0.12	0.99
	20 cm	−2.06	0.14	0.96	5.38	0.24	0.98
	30 cm	16.75	0.14	0.93	8.01	0.34	0.97
	40 cm	18.86	0.24	0.79	14.84	0.42	0.92

REFERENCES

Azam, S., Shah, I., Raghunandan, M.E., et al., 2013. Study on swelling properties of an expansive soil deposit in Saskatchewan, Canada. Bulletin of engineering geology and the environment, 72(1), 25–35.

Bouwer, H., 1969. Infiltration of water into nonuniform soil. Journal of the irrigation and drainage division, 95(4), 451–462.

Bronswijk, J.J.B., 1990. Shrinkage geometry of a heavy clay soil at various stresses. Soil science society of America journal, 54, 1500–1502.

Gan, Y., Jia, Y., Wang, K., et al., 2015. Rainfall infiltration considering gas resistance effect in layered soil. Journal of hydraulic engineering, 46(2), 164–173 (in Chinese).

Green, W.H. and Ampt, G.A., 1911. Studies on soil physics: Part I. Flow of air and water through soils. The journal of agricultural science, 4, 1–24.

Huang, C.Q. and Shao, M.A., 2008. Experimental study on soil shrinking and swelling characteristics during the alternative drying and wetting processes. Chinese Journal of soil science, 39(6), 1243–1247. (in Chinese).

Kim, D.J., Angulo Jaramillo, R., and Vauclin, M., et al., 1999. Modeling of soil deformation and water flow in a swelling soil. Geoderma, 92, 217–238.

Kim, D.J., Diels, J. and Feyen, J., 1992. Water movement associated with overburden potential in a shrinking marine clay soil. Journal of hydrology, 133, 179–200.

Klute, A., 1986. Methods of soil analysis, Part 1: Physical and mineralogical methods, SSSA Book Series 5, 2nd ed., SSSA, Madison, WI.

Lambe, T.W. and Whiman, R.V., 1979. Soil mechanics Six version. 533.

Lockington, D.A. and Parlange, J.Y., 2003. Anomalous water absorption in porous materials. Journal of physics D: Applied physics, 36(6), 760–767.

McGarry, D. and Malafant, K.W.J., 1987. The analysis of volume change in unconfined units of soil. Soil science society of America journal, 51(2), 290–297.

Philip, J.R., 1968. Kinetics of sorption and volume change in clay–colloid pastes. Soil research, 6, 249–267.

Rao, M.D., Raghuwanshi, N.S. and Singh, R., 2006. Development of a physically based 1D-infiltration model for seal formed irrigated soils. Agricultural water management, 84(1–2), 164–174.

Richards, L.A., 1931. Capillary conduction of liquids through porous mediums. Journal of applied physics, 1(5), 318–333.

Su, N., 2014. Mass-time and space-time fractional partial differential equations of water movement in soils: Theoretical framework and application to infiltration. Journal of hydrology, 519, 1792–1803.

Terzaghi, K., 1925. Principles of soil mechanics: a summary of experimental results of clay and sand. English news resources, 3–98.

Van Genuchten, M.T., 1980. A closed-form equation for predicting the hydraulic conductivity of unsaturated soils. Soil science society of America journal, 44(5), 892–898.

Vatankhah, A.R., 2015. Discussion of modified Green–Ampt infiltration model for steady rainfall by J. Almedeij and I.I. Esen. Journal of hydrologic engineering, 20(4), 7014011.

CHAPTER 4

Hydrological Modeling for Karst Structure and Its Application in the Karst Mountain Region

4.1 RESEARCH BACKGROUND

The Karst Mountain Region (KMR), covering 7%–12% of the Earth's continental area, has unique surface and subsurface features compared to the non-karst region (Hartmann et al., 2014). Due to the carbonate nature, karst landscape is generally characterized by thin surface soil, high soil infiltration capacity, and complex topography (Williams, 2008). Hartmann et al. (2014) proposed that several types of porosities exist in karst aquifers: micropores, small fissures, large fractures, and conduits. These types of porosities lead to a strong spatial heterogeneity of hydrogeological conditions. Such particular characteristics play an important role in water and energy balance and affect the dynamics of actual evapotranspiration, infiltration, and runoff in KMR. As pointed out by Kiraly (1998), the hydrological behavior of the karst systems shows a duality in its dynamics: duality of infiltration and recharge processes, duality of the subsurface flow field, and duality of discharge conditions. Because of the complexity of hydrogeological conditions, the simulation and evaluation of hydrological processes in KMR is a challenging and demanding task.

As a powerful tool for modeling hydrological processes and assessing the spatiotemporal variation of water resources, a variety of hydrological models have been developed and applied in karst catchments. In general, these models are distinguished into lumped and distributed karst simulation models. The lumped (or conceptual) models are often used for large-scale river basins, where a karst aquifer is considered as a unit that converts input functions (e.g., precipitation) into output signals (e.g., spring discharge) (Perrin

DOI: 10.1201/9781003646648-4

et al., 2003; Rodríguez et al., 2013). As a result, these models are very limited in describing the spatial variability of hydrological processes in the karst catchments. Contrarily, the distributed hydrological models provide a detailed simulation of the karst systems. The distinct surface and subsurface features in KMR are accounted for by a series of different input parameters in these models (Kordilla et al., 2012; Nikolaidis et al., 2013). However, these parameters show a strong heterogeneity and cannot be applied in medium and large-scale river basins. The Karst Mountain Region of Southwest China (KMRSC) is one of the largest continuous karst areas in the world, which is critical for the water supply of more than 250 million people in the downstream area (Liu et al., 2016). It is estimated that the karst region provides about 54% of the water supply in the catchments (i.e., Wujiang and Xijiang). The majority of the previous studies in the KMRSC have focused on a small catchment scale with an area of several square kilometers for modeling the regional hydrological processes and their impact on water resources. The fracture distribution, pipeline structure characteristics, and groundwater confluence path in the typical catchment are determined by laboratory experiments and field observations in these studies, and then a model with complex structural parameters is established (Chen et al., 2012).

However, the evolution law of the hydrological cycle and water resources in the KMRSC remains unclear from the perspective of sustainable water resources management. Therefore, by selecting the whole KMRSC as the study area, rather than an experimental small catchment, this study attempted to examine the spatiotemporal variation of regional water cycle fluxes. For this purpose, by introducing an equivalent porous medium (EPM) approach into the WEP-L distributed hydrological model, the WEP-karst model was developed for an enhanced description of the karst landform characteristics. Through the EPM approach, the simulation of soil water dynamics for karst hydrogeological characteristics was improved in the WEP-karst model. The special vadose zone in the karst region was divided into four layers: soil, upper epikarst, lower epikarst, and transition layer. Furthermore, the equivalent soil moisture movement parameters of these four layers in the model were proposed and then were approximately estimated based on small-scale experimental observations. Hydrological processes in a river basin are complex and involve a large number of variables. Through the WEP-karst model, more than 20 water cycle fluxes in the KMRSC can be determined. As the key fluxes reflecting regional hydrological dynamics in the vertical direction, precipitation, infiltration, and evapotranspiration were selected for analysis in this study.

Moreover, considering the coordination of water use in production, living, and ecological protection, the blue water and green water were also taken into account in our model. According to the concept proposed by Falkenmark and Rockström (2006), freshwater resources can be classified as blue water or green water. The former is traditionally quantified as the runoff that can be directly used for human consumption, while the latter is the summation of the actual evaporation (the nonproductive part) and the actual transpiration (the productive part) (Rodrigues et al., 2014; Veettil and Mishra, 2016). In some references, only the transpiration is regarded as the green water component (Savenije, 2004; Du et al., 2019). In this study, blue water was defined as the river runoff that provides more than 90% of the water used for production and living in the KMRSC. Green water was defined as vegetation evaporation (i.e., rainwater used efficiently by the vegetation), which is an important indicator of the ecology of mountain vegetation.

4.2 STUDY AREA AND DATA

4.2.1 Study Area

The boundary of the KMRSC was determined based on the South China karst topographic map as well as the Asian karst topographic distribution map (Figure 4.1). The land area is 213,000 km², covering the middle and upper parts of the Pearl River and Yangtze River. In terms of administrative divisions, this area involves the Guizhou and Guangxi provinces, as listed in Table 4.1. The region is characterized by the subtropical wet monsoon climate, with abundant rainfall and high temperatures. The annual average values of precipitation and temperature are about 1500 mm and 17.1°C, respectively. The topography is high in

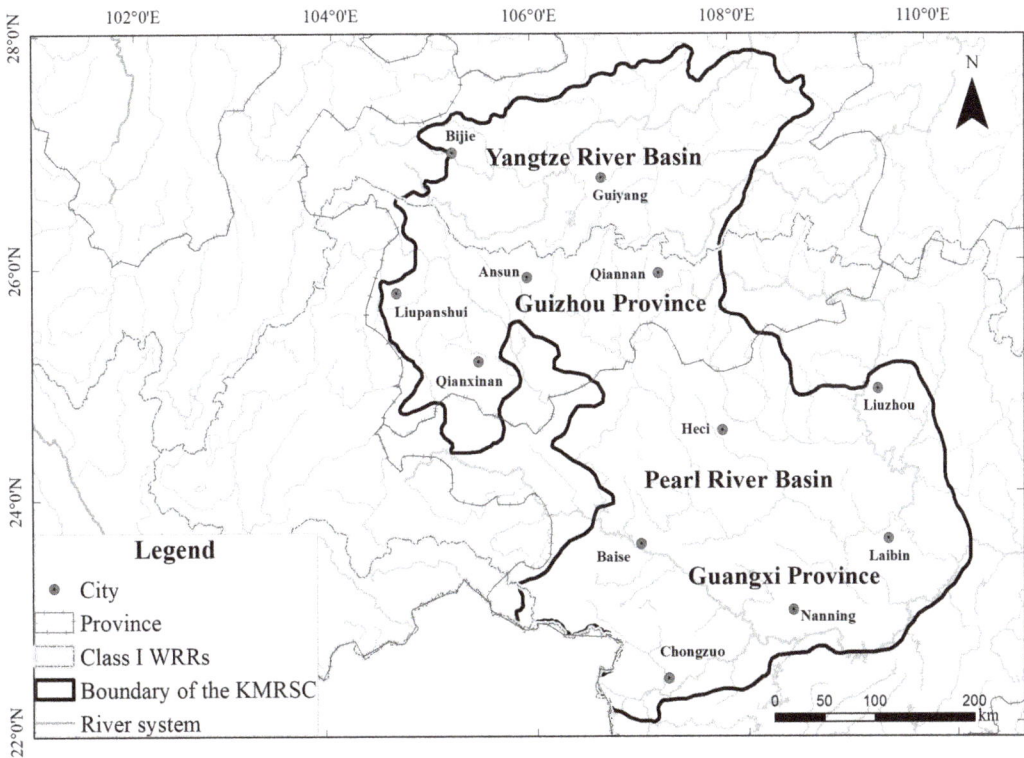

FIGURE 4.1   Geographical location of the KMRSC.

TABLE 4.1   Geographical and Administrative Divisions of the KMRSC

Region	Physical Geographical Divisions			Administrative Divisions		
	Location	Area/10⁴km²	Proportion/%	Location	Area/10⁴km²	Proportion/%
KMRSC	Pearl River	5.3	25	Guizhou	9.5	45
	Yangtze River	16.0	75	Guangxi	11.8	55

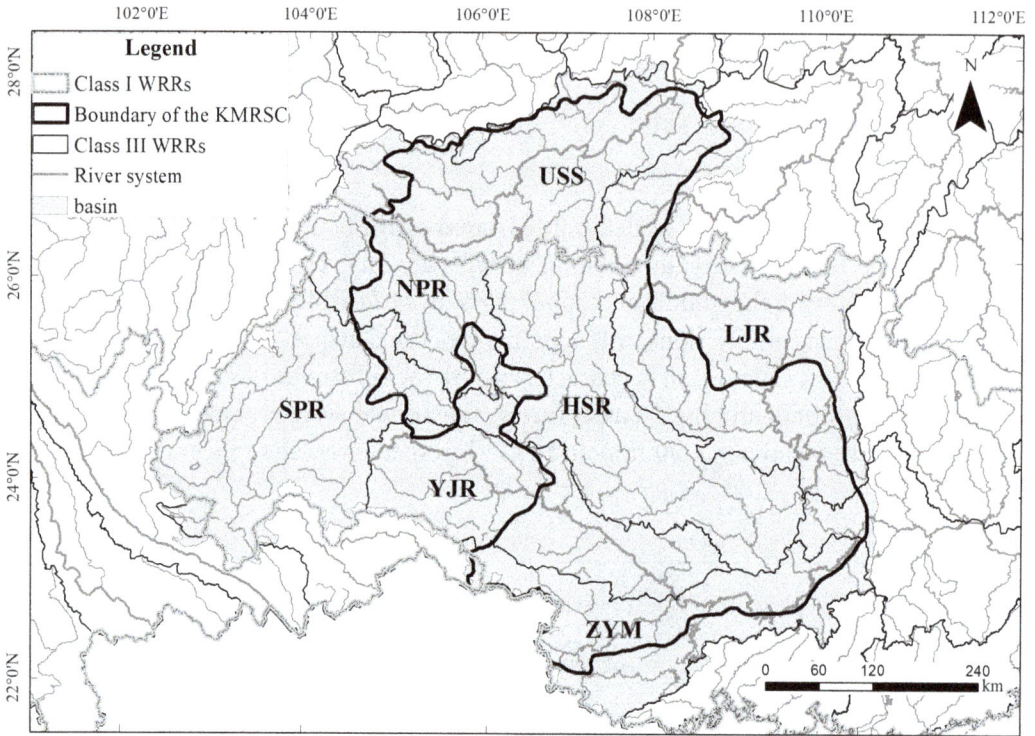

FIGURE 4.2 Water resources regions covered in the modeling domain.

the northwest and low in the southeast. The land-use types in low-, middle-, and high-altitude areas are mainly paddy field and dry land, forest land, and grassland, respectively. The KMRSC involves multiple closed and unclosed basins, while the hydrological model is generally applied to closed basins. Based on the river system involved in the KMRSC, the simulated model domain was appropriately expanded to cover the study area. The modeling domain in this study covered seven of Class III WRRs in China, namely, USS (Upstream Sina Station), SPR (South Panjiang River), NPR (North Panjiang River), HSR (HongShuihe River), LJR (LiuJiang River), YJR (YouJiang River), and ZYM (Zuojiang and Yujiang Main Stream) (Figure 4.2).

4.2.2 Data Preparation

To quantitatively analyze the evolution of regional water cycle fluxes in this study, three types of data needed to be collected and prepared. First, a series of basic data, namely, model input data, was required to run a distributed karst simulation model. The data included the following categories: hydro-meteorology, topography, soil, river system, hydrogeology, land use, and vegetation.

1. Hydro-meteorology data consisted of five items: rain/snow, air temperature, sunshine hours, vapor pressure/relative humidity, and wind speed on a daily basis during the period 1956–2015. 69 key national meteorological stations inside and around

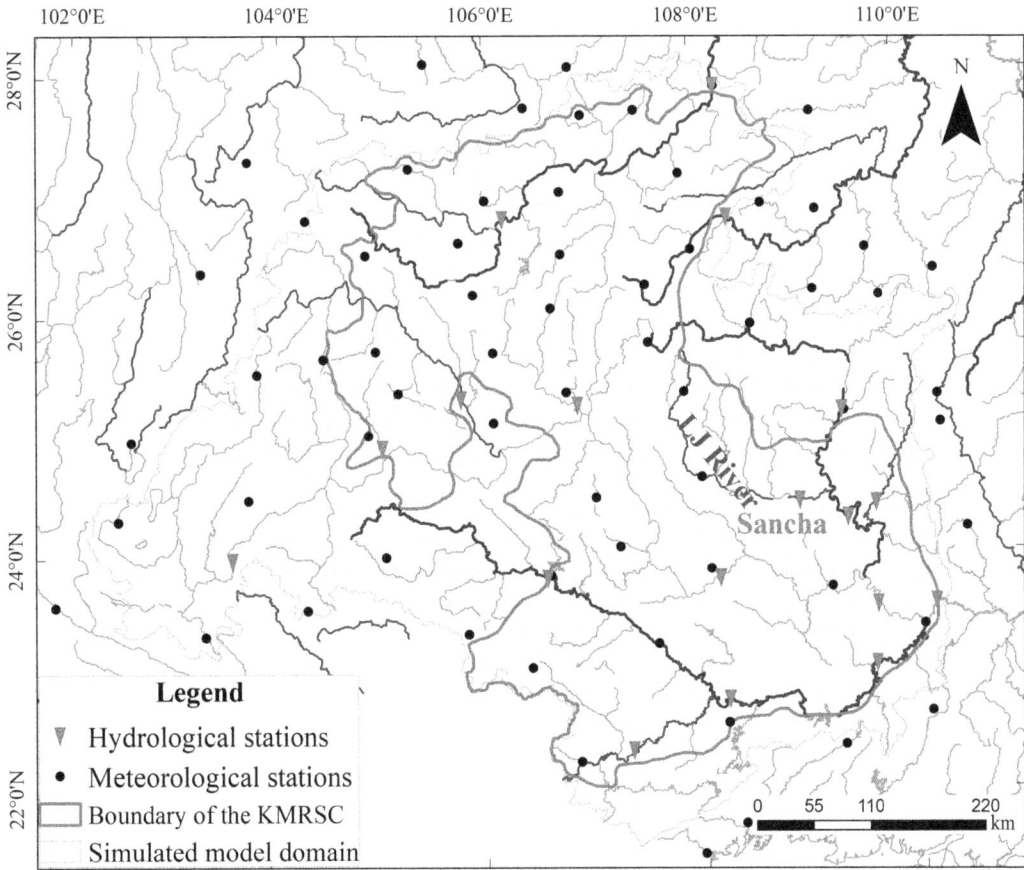

FIGURE 4.3 Location of meteorological and hydrological stations in the modeling domain.

the modeling domain were selected that come from the National Meteorological Information Center of China, as shown in Figure 4.3.

2. Topography data were obtained from the Shuttle Radar Topography Mission (SRTM) with a spatial resolution of 90 m, which can be downloaded from the Geospatial Data Cloud site, Computer Network Information Center, Chinese Academy of Sciences (http://www.gscloud.cn).

3. Soil data, including a map of the 1km grid soil types and corresponding characteristic parameters, were from the National Second Soil Survey Data.

4. River system included a river network map and river cross-sections. The surveyed river network was obtained from the National Geography Database at a scale of 1:250,000. The cross-section information was obtained from the hydrological yearbook.

5. Hydrogeology. The hydrogeology data of aquifers, including hydrogeology unit division, permeability, and storage coefficient, were from the Distribution Map of Hydrogeology in China.

6. The information on land use was the national 1km grid land use type data of five periods (1980, 1990, 1995, 2000, 2005, 2010, and 2015), obtained from the Chinese Academy of Sciences.

7. The monthly vegetation information (LAI and FVC) from 2001 to 2015 was obtained using Moderate Resolution Imaging Spectroradiometer (MODIS) satellite data.

Second, to conduct applicability validation of the model, the monthly discharge data at the main river cross-sections, known as model validation data, were used to be compared with the simulated ones. Because the model simulates natural hydrological processes without consideration of water use, the simulated monthly discharges needed to be compared with the statistical ones, which are the traditional naturalized river discharges by reverting statistical water consumption to the observed ones. To ensure the reliability of the validation data, the results of the second and most recent national water resources survey, completed in 2004, were used. Afterward, we selected 18 representative hydrological stations that are sited throughout the main rivers in the KMRSC and obtained their naturalized monthly discharges from 1956 to 2000.

Third, to assess the impact of climate change on water cycle fluxes in the KMRSC, the simulated data of the regional climate model (RegCM4.0) under the Representative Concentration Pathway (RCP) 4.5 scenarios (the median emission scenarios) were used. Provided by the National Climate Center of China Meteorological Administration, the data included the historical simulation from 1950 to 2005, and the climate change simulation from 2006 to 2099, with a grid spacing of 50 km, and achieved good verification results (Gao et al., 2013).

4.2.3 Mann-Kendall (M-K) Trend Test

We used the Mann-Kendall (M-K) trend test (Kendall, 1975) to quantify the significance of temporal trends for hydrological variables. The M-K trend test is unaffected by the true distribution of the data, unaffected by missing data or irregular spacing of measurements, and less sensitive to outliers. This method has been widely used to evaluate the significance of trends in hydro-meteorological time series. The calculation process is expressed as equations (4.1) and (4.2).

$$S = \sum_{a=1}^{n-1} \sum_{b=a+1}^{n} \mathrm{sgn}(x_b - x_a) \tag{4.1}$$

$$\mathrm{sgn}(x_b - x_a) = \begin{cases} -1, & \text{if} \quad x_b - x_a < 0 \\ 0, & \text{if} \quad x_b - x_a = 0 \\ +1, & \text{if} \quad x_b - x_a > 0 \end{cases} \tag{4.2}$$

where S is the statistic variable of the M-K trend test; n is the number of detected data series; x_a and x_b are the data values in time series a and b ($b > a$), respectively.

$$Z = \begin{cases} \dfrac{S+1}{\sqrt{Var(S)}}, & \text{if } S < 0 \\ 0, & \text{if } S = 0 \\ \dfrac{S-1}{\sqrt{Var(S)}}, & \text{if } S > 0 \end{cases} \tag{4.3}$$

$$Var(S) = \frac{n(n-1)(2n+5) - \sum_{p=1}^{q} t_p(t_p - 1)(2t_p + 5)}{18} \tag{4.4}$$

where Z is the standard normal test statistic. Positive values of Z_S show increasing trends, while negative Z values indicate decreasing trends. t_p denotes the number of ties up to sample p, and q is the number of tied groups. If $|Z| > Z_{1-\alpha/2}$, the null hypothesis is rejected, and the variable exhibits a significant trend at the α level. A two-sided test is used here, so the threshold for the significance test at the 0.05 level is 1.96. In addition, linear trend analysis was also used in the study to complement and contrast the M-K trend analysis.

4.3 DISTRIBUTED HYDROLOGICAL MODELING FOR KARST STRUCTURE

4.3.1 WEP-Karst Model Development

4.3.1.1 Division and Codification of Computation Units

Using the Digital Elevation Model (DEM) with 1 km resolution through resampling, the virtual river networks of the modeling catchment were extracted, and then the sub-basins were divided. As a result, the entire modeling catchment was divided into 2021 sub-basins, 1205 of which were in the KMRSC. The average area of sub-basins is about 170 km², and the maximum and minimum areas are 920 and 55 km², respectively. Moreover, the coincidence degree between the simulated and the actual boundary of the domain reached 99%. Furthermore, each sub-basin was divided into 1-10 contour belts, and thus 12792 contour belts inside small sub-basins were obtained as the computation units of the model. The codification of computation units was carried out based on the stem-branch topological relationship of river networks (Liu et al., 2014).

4.3.1.2 Improvement of Water Movement Simulation in the Karst Vadose Zone

In traditional distributed hydrological models, including the WEP-L model, the vadose zone was simply characterized as a soil-bedrock structure, where soil water movement parameters were only determined according to the texture information of different soil types. However, these parameters fail to reflect soil water dynamics in the special vadose zone in the karst region. Specifically, the karst vadose zone has a special structure that includes soil, epikarst zone, and unsaturated zone (Perrin et al., 2003; Williams, 2008). The uppermost soil layer is usually thin and may even miss. The epikarst zone takes a key role in water storage and movement, as well as for transpiration of vegetation root systems, where the fissure network is strongly developed. Moreover, as the extent and frequency of widening diminishes gradually with depth, epikarst permeability diminishes with depth (Ford and Williams, 2007). The unsaturated zone has poor fissure development, and its water capacity and permeability are significantly reduced compared with those in the epikarst zone. Therefore, the integrated soil-epikarst system mainly controls the infiltration and redistribution of precipitation (Bauer et al., 2005; Aquilina et al., 2006; Bittner et al., 2018).

According to the previous studies, with the increase of basin area, the local influence caused by karst fissure, pipeline and sinkhole is diminishing, and an EPM approach has proved adequate for simulating the karst system in a large river basin (Abusaada and Sauter, 2013; Ghasemizadeh et al., 2015). Therefore, this study attempted to introduce the EPM approach into the existing WEP-L model. In this way, the basic laws obtained from small-scale experimental observations in the KMRSC were effectively generalized, and the WEP-karst model was finally developed.

First, the vadose zone in the KMRSC was divided into four layers: soil, upper epikarst, lower epikarst, and transition layer (see Figure 4.4). The water capacity and permeability of the upper epikarst were lower than those of the lower epikarst, and the transition was the lowest. Second, based on the field investigations of the karst experimental catchment, the equivalent soil moisture movement parameters of the four layers were estimated roughly, as listed in Table 4.2. Investigations by Chen et al. (2012) revealed that, in the KMRSC, the medium-permeability fractures were most widely distributed, and the hydraulic conductivities in the near-surface epikarst layer were of the order 10^{-3} m/s, much larger than

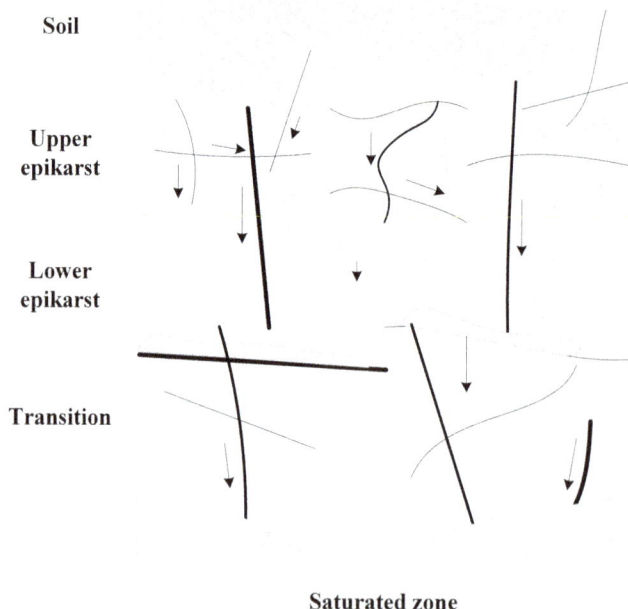

FIGURE 4.4 Schematic representation of vertical hierarchical structure for karst system (modified from Perrin et al., 2003).

TABLE 4.2   Equivalent Soil Moisture Movement Parameters in the Karst System

Parameters		Soil Layer				Epikarst Zone		Transition Layer
Name	Unit	Sand	Loam	Clay Loam	Clay	Upper	Lower	
Soil Thick	cm	50	50	50	50	300	300	200
Soil Porosity	cm³/cm³	0.4	0.466	0.475	0.479	0.12	0.06	0.02
Field Capacity	cm³/cm³	0.174	0.278	0.365	0.387	0.05	0.03	0.01
Residual Moisture Content	cm³/cm³	0.077	0.12	0.17	0.25	0.02	0.01	0.003
Saturated Hydraulic Conductivity	cm/s	2.5E-3	7E-4	2E-4	3E-5	1E-2	1E-3	3E-4

10^{-5} m/s in the low-permeability zones. As for the thickness of the epikarst zone, it varied greatly in space, roughly between 2 and 10m. Finally, the evaporation and transpiration process was further modified, which depends on vegetation species and soil, and rock fractures. According to Xiang et al. (2004), most vegetation roots are concentrated in shallow soil and underlying rock fractures at a depth of less than 1 m. Soil water and epikarst water were the main water sources for plant uptake in the karst forest area (Zhang et al., 2011). Consequently, evaporation of bare soil was assumed to come from the soil layer. Considering root depth, water absorption and transpiration of grass and crop only occurred in the soil and upper epikarst layer, while that of the tree covered the soil and the whole epikarst layer. There was no evapotranspiration in the transition layer because it is difficult for tree roots to extend into this layer due to the hard rock.

4.3.2 Model Calibration and Validation

4.3.2.1 Criteria for Model Calibration and Validation

Using the WEP-karst model, the continuous simulations of 60 years (1956–2015) were conducted in 2021 sub-basins and 12,792 contour belts around the modeling catchment for natural hydrological processes. All of the parameters were initially estimated according to land cover information, observation data, and remote sensing data, and some parameters were selected for model calibration through the sensitivity analysis. The calibration was performed on a basis of "try and error". The calibration parameters included maximum depression storage depth of land surface, soil saturated hydraulic conductivity, hydraulic conductivity of unconfined aquifer, permeability of riverbed material, Manning coefficient, snow melting coefficient, and critical air temperature for snow melting, through the previous sensitivity analysis by Jia et al. (2006). The calibration period was 1956–1980, and the validation period was 1981–2000. Moreover, two widely used criteria for model calibration and validation were used: minimizing the relative error (*RE*) of annually averaged river runoff; maximizing the Nash–Sutcliffe efficiency coefficient (*NSE*) of monthly discharge.

4.3.2.2 Improvement Effect of WEP-Karst Model

Longjiang River basin (LJ) is a typical karst basin in Southwest China, located in the southeast of the study area (Figure 4.3). As the downstream control station of the Longjiang River basin, Sancha station was selected as a representative station to evaluate the improvement effect of water movement simulation in the karst vadose zone. By comparing the simulation results of the WEP-L model and WEP-karst model, it was found that whether the water capacity and permeability of the epikarst zone were considered or not had a great influence

TABLE 4.3   Improvement Effect of WEP-Karst Model at Sancha Station in Longjiang River Basin

Hydrological Stations	WEP Model	Hydrological Process	Periods	NSE	RE
Sancha	Without considering epikarst zone	Monthly discharge	Calibration period (1956–1980)	0.77	10.4%
			Validation period (1981–2000)	0.75	11.5%
	Considering epikarst zone		Calibration period (1956–1980)	0.86	−4.3%
			Validation period (1981–2000)	0.88	−3.8%

on the simulation of the hydrological process in the karst area, as shown in Table 4.3. When the traditional WEP-L model that does not consider the epikarst zone was used, the water storage capacity of the river basin was underestimated, and runoff yield was overestimated. Moreover, the peak discharge was overestimated and the river base flow was underestimated, which resulted in a steeper monthly discharge hydrograph. By considering the epikarst zone, the monthly discharge hydrograph simulated by the improved WEP-karst model was more consistent with the statistical result, with a higher *NSE* and lower *RE*.

4.3.2.3 Model Applicability in the Karst Mountain Region of Southwest China

Calibration and validation results of simulated monthly discharge through the WEP-karst model at the 18 representative hydrological stations are shown in Table 4.4 and Figure 4.5. For the calibration period, the *NSE* values were in the range 0.73–0.94, and the *RE* values

TABLE 4.4   Calibration and Validation Results of Simulated Monthly Discharge at Representative Hydrological Stations

Hydrological Stations	Calibration Period (1956–1980)		Validation Period (1981–2000)	
	NSE	RE	NSE	RE
Yachihe	0.81	−1.8%	0.86	2.6%
Sinan	0.86	8.5%	0.90	4.1%
Shidong	0.80	2.1%	0.76	3.1%
Jiangbianjie	0.73	−6.4%	0.71	−9.8%
Zhexiang	0.84	2.4%	0.80	−6.0%
Zhedong	0.74	4.8%	0.78	0.4%
Bamao	0.81	0.4%	0.74	−2.4%
Baise	0.8	8.3%	0.76	7.6%
Duan	0.81	−2.4%	0.80	8.3%
Chongzuo	0.84	−0.8%	0.91	2.6%
Nanning	0.91	2.6%	0.94	1.4%
Guigang	0.83	4.1%	0.81	4.7%
Changan	0.94	1.4%	0.87	−3.5%
Sancha	0.86	−4.3%	0.88	−3.8%
Liuzhou	0.94	−3.6%	0.93	−2.0%
Duiting	0.92	1.2%	0.93	−3.3%
Wuxuan	0.86	3.8%	0.83	4.5%
Dahuang	0.85	3.9%	0.81	6.5%

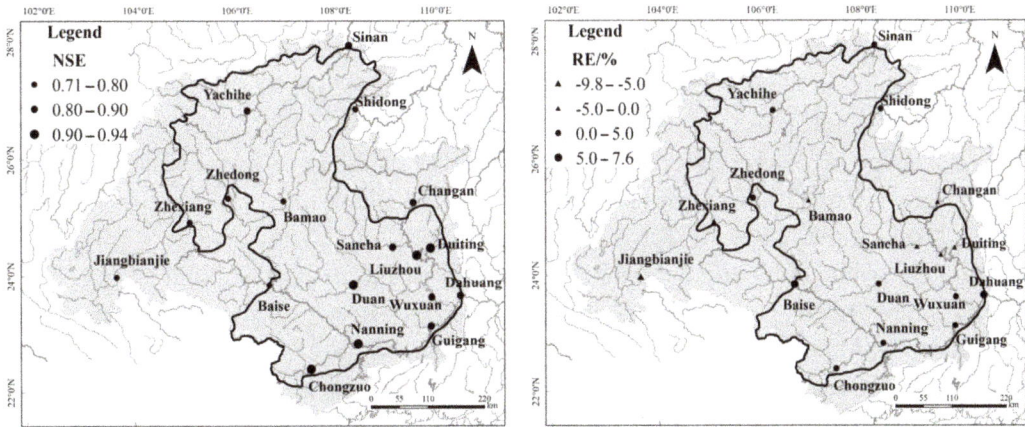

FIGURE 4.5   Schematic illustration of model validation results: (a) NSE, (b) RE.

ranged from −6.4% to 8.5%. There were 14 stations with the *NSE* values larger than 0.8 and the absolute values of *RE* less than 5%, accounting for 78% of the total number of stations. Among all stations, only Jiangbianjie and Zhexiang stations have the *NSE* values lower than 0.8, and Sinan, Jiangbianjie, and Baise stations have the absolute value of *RE* larger than 5%. Similarly, for most stations, the *NSE* values were mostly above 0.8, and the *RE* values were controlled between −5% and 5% during the validation period. In general, both the *NSE* and *RE* values were quite encouraging, which showed that the WEP-karst model has a good applicability across the KMRSC.

4.4 QUANTITATIVE ASSESSMENT OF MAIN WATER CYCLE FLUXES

4.4.1 Vertical Water Cycle Fluxes

The annual average precipitation, infiltration, and evapotranspiration from 1956 to 2015 were determined in all 2021 sub-basins using the WEP-karst model. Furthermore, the amounts were aggregated to the seven Class III WRRs and the KMRSC, as shown in Table 4.5. The annual average precipitation in the modeling domain ranged from 900 to 2240 mm and gradually increased from northwest to southeast, affected by the southwest

TABLE 4.5   Annual Average Precipitation, Infiltration, and Evapotranspiration in the Class III WRRs and the KMRSC

Regions	Precipitation (mm)	Infiltration (mm)	Evapotranspiration (mm)
USS	1265	749	732
SPR	1217	857	880
NPR	1408	827	856
HSR	1593	954	995
LJR	1659	823	833
YJR	1446	892	1085
ZYM	1597	859	904
KMRSC	1506	862	870

and southeast monsoons. Moreover, the annual average precipitation of less than 1300 mm formed a belt distribution in the northwest. There existed three high precipitation centers larger than 1800 mm, concentrated in the east and south. For the seven Class III WRRs, the LJR had the largest precipitation, followed by the ZYM and HSR, with the annual average of nearly 1600 mm and more. In the USS and SPR, the values were the lowest, only about 1200 mm. In terms of the KMRSC, the annual average precipitation (1956–2015) reached 1506 mm, about 2.4 times the national average.

From the whole modeling domain, the spatial distribution of the annual average infiltration was mainly reflected in the difference between the upstream and downstream, with a range of 293–1934 mm. In most river basins, the amount of infiltration in the upper reaches was generally smaller than that in the lower reaches. The reason may be that the thickness of the vadose zone in the upper reaches was relatively small, resulting in a smaller soil water storage capacity and infiltration. Moreover, the closer the area to the main channel, the greater the infiltration values. There was no significant difference in the annual average infiltration among these Class III WRRs. The infiltration amount in the USS was the smallest, 749 mm, while that in the HSR was the largest, 954 mm. The KMRSC had an annual average infiltration of 862 mm, which accounted for 57% of the precipitation.

Affected by meteorological, hydrological, vegetation, and soil conditions, the annual average evapotranspiration generally showed a spatial distribution pattern of high in the south and low in the north. The evapotranspiration exhibited large spatial variability in the modeling domain, varying from 550 to 1300 mm. The amount was the lowest in the USS located in the Yangtze River Basin, only 732 mm. However, in the HSR and YJR, the values reached about 1000 mm. There existed some areas with high evaporation value in the SPR, LJR, HSR, and YJR. For the KMRSC, the annual average evapotranspiration was 870 mm, roughly equivalent to the infiltration.

4.4.2 Blue Water and Green Water

The annual average blue water showed a similar spatial distribution with the precipitation from northwest to southeast, which had a range of 100–1500 mm (Figure 4.6). The areas with abundant blue water resources were concentrated in the east of the LJR, the middle of the HSR, and the northwest of the ZYM. The river runoff yield in the upper reaches of the SPR was relatively small and insufficient, less than 400 mm. At the Class III WRRs scale, the amount of blue water was the largest in the LJR with 918 mm, which was significantly higher than that in other regions. In contrast, the USS, YJR, and SPR had low availability of blue water resources, which varied from 417 to 560 mm. In the KMRSC, the annual average blue water reached 701 mm, which means that the local per capita river runoff was about 3100 m³.

Green water resource is an important part of regional evapotranspiration, especially in areas with high precipitation and dense vegetation cover. In the modeling domain, the annual average green water was in a range of 36–886 mm, and its spatial variation seemed to be consistent with that of the evapotranspiration and the NDVI, as shown in Figure 4.7. Affected by the precipitation and vegetation conditions, there existed some differences in the annual average green water and its proportion in the evapotranspiration among the

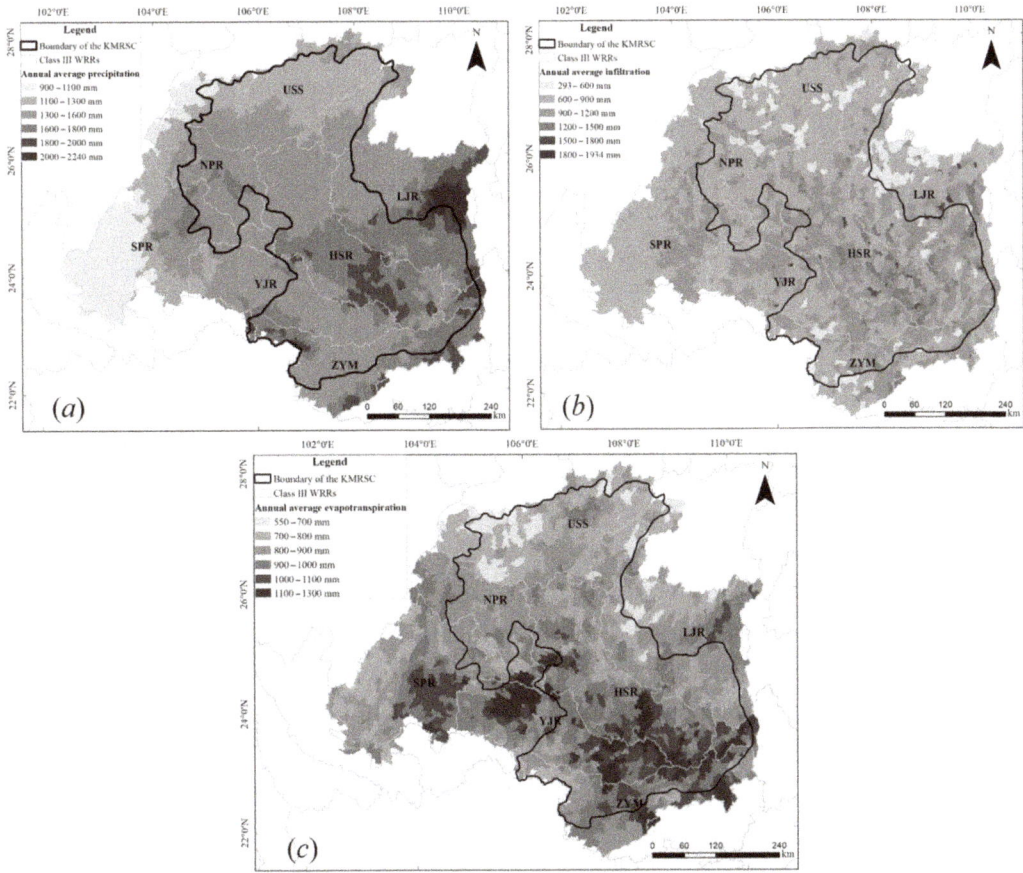

FIGURE 4.6   Spatial distribution of annual average blue water in the modeling domain.

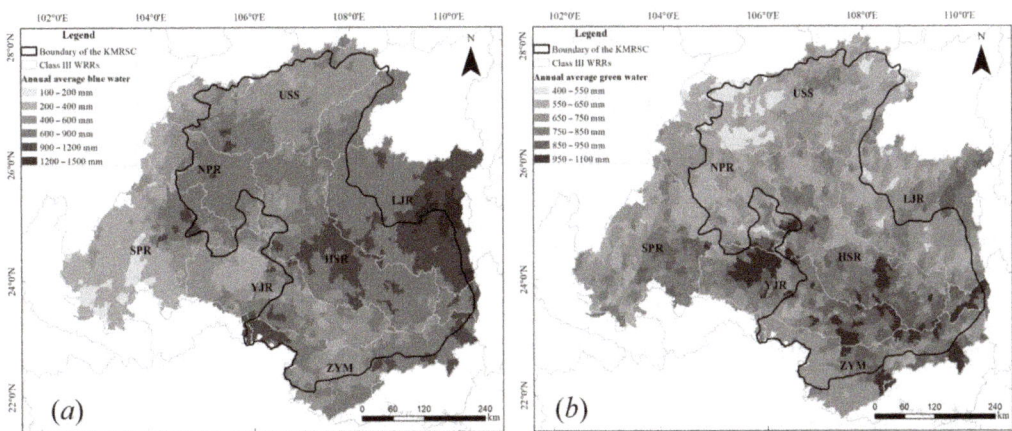

FIGURE 4.7   Spatial distribution of annual average green water in the modeling domain.

TABLE 4.6   Annual Average Blue Water and Green Water in the Class III WRRs and the KMRSC

Regions	Blue Water (mm)	Green Water	
		Amount (mm)	Proportion (%)
USS	560	388	53.01
SPR	417	446	50.68
NPR	616	346	40.42
HSR	741	495	49.75
LJR	918	479	57.50
YJR	532	562	51.80
ZYM	694	477	52.77
KMRSC	701	445	51.15

Note: Amount and Proportion represent the annual average green water and its proportion in the evapotranspiration, respectively.

seven Class III WRRs, as listed in Table 4.6. The amount in the YJR was the largest, with 562 mm, while in the NPR, the amount was only 346 mm. Moreover, the green water resources in the USS and NPR were similar, but their proportions were significantly different. It may be mainly due to the difference in precipitation between the two regions.

By comparing the spatial distribution of blue and green water, some interesting information was found. First, in the YJR and the SPR, the green water resource was relatively high, while the amount of blue water was low, which may be related to the vegetation's capacity to hold soil moisture and reduce runoff. Especially in the upper reaches of the YJR, most of the precipitation was consumed by vegetation transpiration, so the reasonable vegetation coverage threshold needs to be further explored. Second, in the southwest of the KMRSC, there were low green water resources and high river runoff. Due to the poor vegetation conditions, the area faces a relatively high flood risk. Third, in some areas, the precipitation was large, and the vegetation conditions were good, so both the blue water and green water were relatively high, such as in the east of the LJR and the lower reaches of the HSR. In this case, the balance between water uses in industrial and agricultural production and ecology needs to be focused on.

4.5 DYNAMICS OF MAIN WATER CYCLE FLUXES

4.5.1 Annual Variability of Main Water Cycle Fluxes

Focused on the KMRSC, the annual change trends of water cycle fluxes from 1956 to 2015 were analyzed through the nonparametric M-K test. Besides, the coefficient of variation (CV) was adopted to analyze the annual fluctuation of these fluxes. It can be seen from Table 4.7 and Figure 4.8 that all fluxes showed a decreasing trend, of which the annual precipitation, infiltration, and evapotranspiration decreased at the $p = 0.01$ significant level with a change rate of −2.59, −1.48, and −1.65 mm/year, respectively. The amount of blue water resource decreased at the $p = 0.05$ significant level with a change rate of −2.07 mm/year, while green water indicated an insignificant trend. It should be noted that due to extreme drought events in southwest China around 2010, precipitation and river runoff

TABLE 4.7   Mann-Kendall Test Results and the Values of CV for Water Cycle Fluxes in the KMRSC

Water Cycle Fluxes	Mann-Kendall Test			Coefficient of Variation (CV)
	Z statistic	Significance	Change Rate (mm/year)	
Precipitation	−2.9	**	−2.59	0.13
Infiltration	−2.7	**	−1.48	0.12
Evapotranspiration	−3.4	**	−1.65	0.08
Blue Water	−2.1	*	−2.07	0.28
Green Water	−0.8	–	−0.14	0.07

Note: ** and * indicate significance levels of $p = 0.01$ and $p = 0.05$, respectively; – means $p > 0.05$.

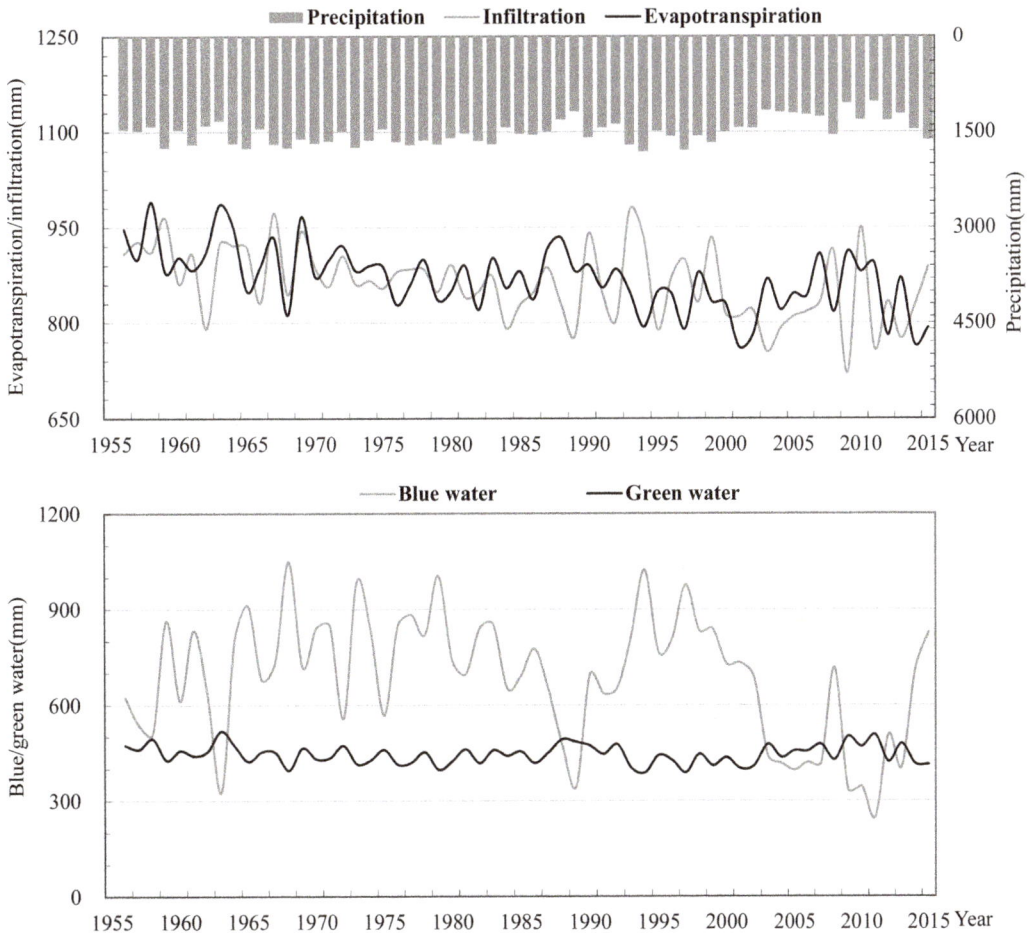

FIGURE 4.8   Annual variability of water cycle fluxes in the KMRSC.

decreased to historically low levels. Furthermore, there existed significant differences in the annual fluctuations of the five fluxes. In terms of the annual blue water, the CV value reached 0.28, and the fluctuation range was large, which brings more uncertainty to the regional water supply. The annual fluctuation of the precipitation was close to the average level in southern China, with the CV value of 0.13. In contrast, the annual infiltration showed a similar fluctuation range to the precipitation, while the evapotranspiration and green water had a weak fluctuation. This may be due to the rich and stable groundwater in the karst region, resulting in the limited impact of annual fluctuations in precipitation and runoff on evapotranspiration and green water.

4.5.2 Intraannual Variability of Main Water Cycle Fluxes

The monthly average values of water cycle fluxes in the KMRSC are listed in Table 4.8. Moreover, the box diagram was chosen presentation of the intraannual variability of water cycle fluxes in the KMRSC, as shown in Figure 4.9. It reflected not only the monthly distribution of these fluxes but also the change of monthly data over the past 60 years. A fast increase in precipitation was recorded in April and May. The highest monthly precipitation was 273 mm in June, while the lowest was 36 mm in January. The precipitation concentrated from May to August, accounting for 62.2% of the annual total precipitation. Besides, during the rainy period (June to August), the precipitation showed a relatively strong fluctuation. In other dry months, the fluctuation of precipitation in October was the strongest. Like the precipitation, 58.6% of the annual total infiltration occurs from May to August. However, the fluctuation range of infiltration was relatively large in each month. Compared with the precipitation and infiltration, evapotranspiration seemed to be more evenly distributed on the monthly scale, and the fluctuation of monthly data was smaller. The maximum value of monthly evaporation appeared in July with 129 mm, which was one month later than that of precipitation and infiltration. Blue water increased strongly from May to June and peaked in June with 136 mm. The values in the first three months of a year were the smallest, accounting for only 6.6% of the annual blue water resource. Interestingly, the larger the monthly blue water, the stronger the fluctuation seems. Due to the close relationship with vegetation growth, the distribution difference of green water in each month was obvious. The amounts in July and August were significantly higher than those in other months, accounting for 38.9% of the total annual value. The fluctuation range of green water in the months between May and September was larger than that in other months.

TABLE 4.8   Monthly Average Values (mm) of Water Cycle Fluxes in the KMRSC

Month	Precipitation	Infiltration	Evapotranspiration	Blue Water	Green Water
Jan	36	25	34	15	7
Feb	39	27	40	14	8
Mar	59	41	59	17	16
Apr	107	70	79	27	30
May	209	127	98	66	48
Jun	273	141	101	136	59
Jul	246	128	129	134	89
Aug	209	109	115	113	84
Sep	132	78	84	72	50
Oct	99	59	60	51	32
Nov	60	35	38	36	13
Dec	37	23	33	21	8
Total	1506	862	870	701	445

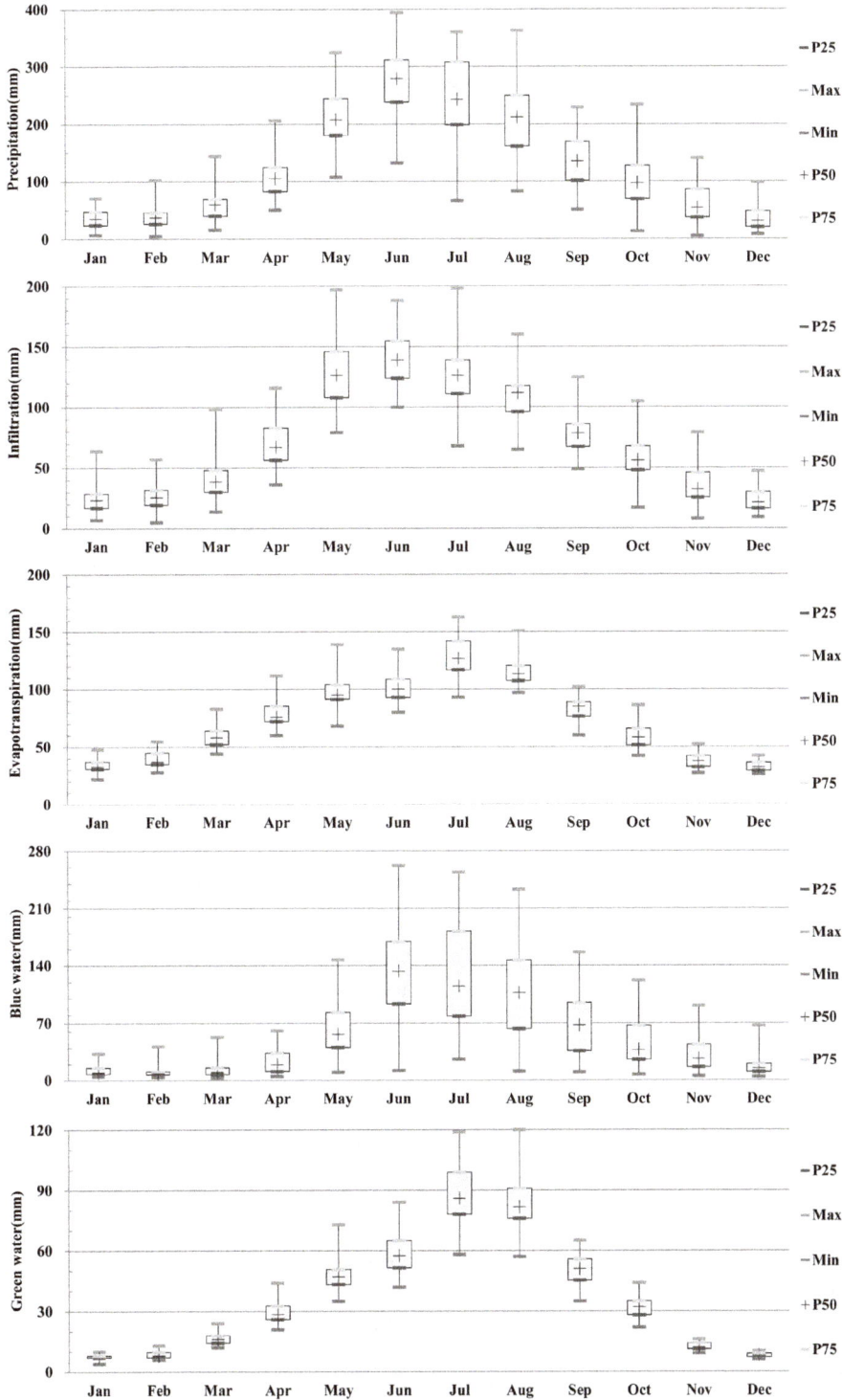

FIGURE 4.9   Intraannual variability of water cycle fluxes in the KMRSC. Min, P25, P50, P75, and max represent the minimum, lower quartile, median, upper quartile, and maximum of monthly data series from 1956 to 2015.

TABLE 4.9 Annual Average Amounts of Water Cycle Fluxes during the Prediction Period and Their Change Rates Compared to the Base Period Results

	Precipitation		Infiltration		Evapotranspiration		Blue Water		Green Water	
Regions	**Amount/mm**	**Rate/%**	**Amount/mm**	**Rate/%**	**Amount/mm**	**Rate/%**	**Amount/mm**	**Rate/%**	**Amount/mm**	**Rate/%**
USS	1218	−3.7	746	−0.4	746	1.9	471	−16.0	394	1.4
SPR	1160	−4.7	845	−1.4	873	−0.8	349	−16.3	444	−0.5
NPR	1360	−3.4	825	−0.2	860	0.5	518	−15.8	347	0.4
HSR	1542	−3.2	925	−3.0	961	−3.4	648	−12.5	481	−2.8
LJR	1616	−2.6	822	−0.1	840	0.8	819	−10. 8	481	0.4
YJR	1428	−1.2	891	−0.2	1061	−2.2	493	−7.2	555	−1.3
ZYM	1605	0.5	860	0.1	930	2.9	680	−2.0	485	1.6
KMRSC	1461	−3.0	858	−0.46	875	0.6	625	−10.8	447	0. 5

4.5.3 Assessing the Impact of Climate Change

Considering the rough prediction data, the impact of climate change on water cycle fluxes was assessed and discussed at the Class III WRRs scale through the WEP-karst model, as shown in Table 4.9. In this study, the period 1956–2015 was set as the base period, and the period 2021–2050 was set as the prediction period. For the KMRSC, a clear trend with the decrease in precipitation and blue water resource under the median emission scenario (RCP4.5) was observed, with the change rate of −3.0% and −10.8%, respectively. The infiltration, evapotranspiration, and green water changed slightly compared to the base period results. Among the Class III WRRs, the change rate of annual average precipitation varied between −4.7% and 0.5%. Spatially, the river runoff yield in the USS, SPR, and NPR, located in the upper reaches of the KMRSC, seemed to be more sensitive to climate change and had a more dramatic decrease, with the change rate of about −16%. As a result, increasing water scarcity risk is expected in the KMRSC. In terms of the annual average infiltration, evapotranspiration, and green water, the change in the HSR was larger than that in other regions. It may be that the HSR is in the downstream area, with good vegetation conditions and a thick soil layer, so it is greatly affected by the decrease in precipitation and the increase in temperature.

REFERENCES

Abusaada, M. and Sauter, M., 2013. Studying the flow dynamics of a karst aquifer system with an equivalent porous medium model. Groundwater, 51(4), 641–650.

Aquilina, L., Ladouche, B. and Doerfliger, N., 2006. Water storage and transfer in the epikarst of karstic systems during high flow periods. Journal of hydrology, 327, 472–485.

Bauer, S., Liedl, R. and Sauter, M., 2005. Modeling the influence of epikarst evolution on karst aquifer genesis: A time-variant recharge boundary condition for joint karst-epikarst development. Water resources research, 41, W0941.

Bittner, D., Narany, T.S., Kohl, B., et al., 2018. Modeling the hydrological impact of land use change in a dolomite-dominated karst system. Journal of hydrology, 567, 267–279.

Chen, X., Zhang, Y.F., Xue, X., et al., 2012. Estimation of baseflow recession constants and effective hydraulic parameters in the karst basins of southwest China. Hydrology research, 43(1–2), 102–112.

Du, J., Jia, Y. and Hao, C. et al., 2019. Temporal and spatial changes of blue water and green water in the Taihang Mountain Region, China, in the past 60 years. Hydrological sciences journal, 64(16), 2040–2056.

Falkenmark, M. and Rockström, J., 2006. The new blue and green water paradigm: Breaking new ground for water resources planning and management. Journal of water resources planning and management, 132(3), 129–132.

Ford, D.C. and Williams, P.W., 2007. Karst Hydrogeology and Geomorphology. Wiley, Chichester, 561.

Gao, J., Wang, M. and Giorgi, F., 2013. Climate change over China in the 21st century as simulated by BCC_CSM1. 1-RegCM4. 0. Atmospheric and oceanic science letters, 6(5), 381–386.

Ghasemizadeh, R., Yu, X., Butscher, C., et al., 2015. Equivalent porous media (EPM) simulation of groundwater hydraulics and contaminant transport in karst aquifers. Plos one, 10(9), e0138954.

Hartmann, A., Goldscheider, N., Wagener, T., et al., 2014. Karst water resources in a changing world: Review of hydrological modeling approaches. Reviews of geophysics, 52(3), 218–242.

Jia, Y., Wang, H., Zhou, Z., et al., 2006. Development of the WEP-L distributed hydrological model and dynamic assessment of water resources in the Yellow River basin. Journal of hydrology, 331(3–4), 606–629.

Kendall, M.G., 1975. Rank Correlation Measures. Charles Griffin and Company, Ltd, London, UK, 701, 1–202.

Kiraly, L., 1998. Modelling karst aquifers by the combined discrete channel and continuum approach. Bulletin d'hydrogéologie, 16, 77–98.

Kordilla, J., Sauter, M., Reimann, T., et al., 2012. Simulation of saturated and unsaturated flow in karst systems at catchment scale using a double continuum approach. Hydrology and earth system sciences, 16(10), 3909–3923.

Liu, J., Zhou, Z., Jia, Y., et al., 2014. A stem-branch topological codification for watershed subdivision and identification to support distributed hydrological modelling at large river basins. Hydrological processes, 28(4), 2074–2081.

Liu, M., Xu, X., Wang, D., et al., 2016. Karst catchments exhibited higher degradation stress from climate change than the non-karst catchments in southwest China: An eco-hydrological perspective. Journal of hydrology, 535, 173–180.

Nikolaidis, N.P., Bouraoui, F. and Bidoglio, G., 2013. Hydrologic and geochemical modeling of a karstic Mediterranean watershed. Journal of hydrology, 477, 129–138.

Perrin, J., Jeannin, P.Y. and Zwahlen, F., 2003. Epikarst storage in a karst aquifer: A conceptual model based on isotopic data, Milandre test site. Switzerland. Journal of hydrology, 279(1–4), 106–124.

Rodrigues, D.B., Gupta, H.V. and Mendiondo, E.M., 2014. A blue/green water-based accounting framework for assessment of water security. Water resources research, 50, 7187–7205.

Rodríguez, L., Vives, L. and Gomez, A., 2013. Conceptual and numerical modeling approach of the Guarani Aquifer System. Hydrology and earth system sciences, 17(1), 295–314.

Savenije, H.H.G., 2004. The importance of interception and why we should delete the term evapotranspiration from our vocabulary. Hydrological processes, 18, 1507–1511.

Veettil, A.V. and Mishra, A.K., 2016. Water security assessment using blue and green water footprint concepts. Journal of hydrology, 542, 589–602.

Williams, P.W., 2008. The role of the Epikarst in karst and cave hydrogeology: A review. International journal of speleology, 37(1), 1–10.

Xiang, C., Song, L., Zhang, P., et al., 2004. Preliminary study on soil fauna diversity in different vegetation cover in Shilin National Park, Yunnan, China. Resources science, 26, 98–103 (in Chinese).

Zhang, Z.C., Chen, X., Ghadouani, A., et al., 2011. Modeling hydrological processes influenced by soil, rock and vegetation in a small karst basin of southwest China. Hydrological processes, 25(15), 2456–2470.

CHAPTER 5

Hydrothermal Coupling Modeling for the Cold Region and Its Application in the Source Area of the Yangtze River

5.1 RESEARCH BACKGROUND

Currently, research on the cold regions hydrology mainly focuses on glacier meltwater, snowmelt runoff, permafrost processes, and high-latitude freshwater balance, revealing the synergistic changes in water and energy and their impacts on water security through observation and simulation methods (Ding et al., 2020). Significant research achievements have been made both domestically and internationally in the simulation of glacier, snow, and permafrost hydrology. Among them, temperature index models and energy balance models are two basic methods for simulating the meltwater runoff processes of glaciers and snow. The former relies on the correlation between temperature and the amount of glacier or snow melt, with a relatively simple structure and inputs, which is suitable for the large-scale catchment. Representative models include the early developed SRM model (Ma, 2003) and the widely used SPHY model in the Third Pole (Terink et al., 2015). Compared to temperature index models, energy balance models have more sophisticated physical mechanisms. Commonly used models include SNOBAL (Zhou et al., 2021a) and UEB (Liu et al., 2020). However, some parameters in these models require locally fine-grained observations or experiments, so this type of model is mostly used at point scales or in small catchments. Ding et al. (2017) developed the enthalpy-based glacier mass and energy balance model, WEB-GM, which has been successfully applied to the ablation area of the Parlung No. 4 Glacier on the Qinghai-Tibet Plateau. Regarding permafrost hydrological processes,

DOI: 10.1201/9781003646648-5

relevant scholars have developed one-dimensional permafrost hydrothermal coupling models, among which SHAW (Flerchinger and Saxton, 1989a) and CoupModel (Jansson and Moon, 2001) are the most representative. These models can simulate the entire process of permafrost freezing and thawing, but they suffer from issues such as numerous parameters, difficult parameter acquisition, and slow computation speed, which limit their application in large-scale catchments.

To meet the research demands of large-scale catchment runoff variations in the cold regions, some scholars are committed to incorporating the hydrological processes of glaciers, snow, and permafrost changes in the traditional distributed hydrological models. Currently, models such as WEP-L (Jia et al., 2006), SWT (Fontaine et al., 2002), and GBEHM (Gao et al., 2018) have considered the influence of glaciers/snow and permafrost on runoff generation, but their process representation is weak. For example, based on the WEP-L model, scholars have developed the WEP-COR model (Li et al., 2019a), considering permafrost freezing and thawing processes, where a hydrothermal coupling module is used to calculate soil water and heat transfer. Based on the WEP-COR model, Zhou et al. (2021b) developed a hydrological model considering snow-permafrost-gravel. In the above models, simple empirical formulas were used for the linking equations and characteristic parameters of permafrost hydrothermal coupling, so the formulas and their parameters lacked physical significance, which limited the applicability of the models. Overall, the representation of glacier/snow meltwater and permafrost runoff processes in large-scale distributed hydrological models is not sufficiently comprehensive.

Against the background of climate change, the elements of the cryosphere have a significant impact on regional hydrological processes in the cold regions. This chapter attempted to make two model improvements based on WEP-L for the widely distributed glaciers, snow, and permafrost in the cold regions. On the one hand, the temperature index model (i.e., degree-day factor method) was used to calculate the meltwater processes of glaciers and snow. On the other hand, with reference to the permafrost freezing and thawing module of the one-dimensional permafrost hydrothermal coupling model, the relationship between soil water potential and temperature, and the permafrost freezing and thawing parameterization scheme were introduced to realize the scientific description of the permafrost runoff process in a large-scale catchment. Taking the source area of the Yangtze River (SAYR) as the study area, a distributed hydrothermal coupling model named WEP-SAYR was developed, which comprehensively considers glacier and snow melt and permafrost runoff processes. Model applicability is verified using multiple elements, including runoff, evapotranspiration, and soil temperature and moisture.

5.2 CONSTRUCTION OF THE DISTRIBUTED HYDROTHERMAL COUPLING MODEL

5.2.1 Calculation of Water and Energy Exchange in Glacier and Snow

The degree-day factor method was used to calculate glacier and snow meltwater processes. The study used 1980 as the initial simulation year. Therefore, the initial information on glacier distribution, area, and ice reserves was obtained from the first Chinese Glacier

Inventory, which was completed from 1956 to 1983. These data were then converted to contour belts and participated in the calculation of water and heat fluxes as a new land use type. The meltwater from glaciers and snow was calculated as equation (5.1):

$$M = \begin{cases} DDF \cdot (T_a - T_0) & T_a > T_0 \\ 0 & T_a \leq T_0 \end{cases} \tag{5.1}$$

where M represents the daily melting amount, mm/d; DDF stands for the degree-day factor of glaciers and snow, mm/(°C/d), indicating the amount of glacier and snow melt produced per unit of positive temperature accumulation; T_a is the temperature, °C; T_0 is the critical melting temperature, °C.

The degree-day factors of glaciers and snow were found to be different. According to Zhang et al. (2019a), the degree-day factors of glaciers and snow were calculated using latitude (Lat, °), longitude (Lon, °), glacier terminus altitude (H, m), annual average temperature (T, °C), and precipitation (P, mm).

The equation for atmospheric-land energy exchange in glacier regions was equation (5.2):

$$Q_{net} = (R + H - LE + Q_p + Q_g)\Delta t \tag{5.2}$$

where Q_{net} represents the net energy input to the glacier/snow layer, J; R is the net radiation flux, H is the sensible heat flux, LE is the latent heat flux, Q_p is the heat flux from precipitation input, and Q_g is the heat flux below the glacier, W/m².

The heat flux was calculated by the following equation (Liu et al., 2004):

$$Q_g = \lambda_i (T_f - T_s)/h \tag{5.3}$$

where λ_i is the thermal conductivity of ice, W/(cm/K); T_f is the freezing point, °C; T_s is the surface temperature of the ice, generally substituted by air temperature, °C; h is the thickness of the ice, m. Considering the considerable thickness of glaciers, the study assumes a value of 10 m.

The formula for calculating the heat transfer between the snow and the soil was equation (5.4):

$$Q_g = \frac{1}{\dfrac{Z_s}{2\lambda_s} + \dfrac{Z_c}{2\lambda_c} + R_c}(T_c - T_s) \tag{5.4}$$

where Z_s is the thickness of the snow layer, cm; Z_c is the thickness of the surface soil layer, cm; λ_s is the thermal conductivity of the snow layer, W/(m/°C); λ_c is the thermal conductivity of the soil, W/(m/°C); R_c is the contact thermal resistance between the soil layer and the snow layer, (m²/°C)/W; T_c is the temperature of the surface soil, °C; T_s is the temperature of the snow layer, °C.

When calculating with the model, if there is snow cover on the glacier, calculate the snowmelt first. After the snow cover on the glacier has completely melted, proceed with the calculation of glacier meltwater.

5.2.2 Simulation of Permafrost Hydrothermal Transfer Processes

By analyzing the characteristics of existing hydrological models that consider permafrost simulation, it was found that the WEP-COR model and the CoLM model better consider the inconsistency between hydrological simulation and the depiction of the hydrothermal transfer process of permafrost on the spatial scale. However, the mathematical description of the mechanism of change of several hydrothermal parameters is discarded and replaced by empirical formulas, as shown in Table 5.1. The SHAW model is complicated for the calculation of permafrost hydrothermal processes, which makes it difficult to be directly applied to large-scale catchments. This study is based on the WEP-L model and draws on the advantages of the WEP-COR model, CoLM model, and SHAW model in permafrost hydrothermal transfer processes to further optimize the depiction of the mechanism behind changes in hydrothermal parameters.

To improve the accuracy of the energy balance calculation, the permafrost layer was subdivided into 11 layers (originally 3 layers in WEP-L), taking into account the model

TABLE 5.1 Characteristics of Existing Permafrost Simulation Models

Model	Authors	Characteristics	Permafrost Simulation
GBHM WEP-L	Gao et al. (2018) Jia et al. (2006)	Large-scale hydrological model; Ignoring the physical processes of permafrost freeze-thaw	The calculation is simple but only establishes an empirical relationship between permafrost soil hydraulic conductivity and decreasing air temperature, without considering changes in soil moisture.
WEP-COR	Li et al. (2019); Zhou et al. (2021b)	Large-scale hydrothermal coupled model	The process of permafrost freezing and thawing is well characterized, but the variation of soil hydrothermal parameters is simple to characterize, and the empirical equations of variation with temperature are mostly used.
SWAT	Fontaine et al. (2002)	Energy is modeled using a heat transfer model, and water movement is modeled using the water balance equation.	The model takes into account the heat transfer process but does not effectively portray the physical process of permafrost moisture movement.
CoLM	Liu et al. (2015); Dai et al. (2003)	Large-scale hydrothermal coupled model	Substrate potential and liquid water content are calculated using the Clapp-Hornberger model, which may underestimate the minimum liquid water content.
SHAW CoupModel	Flerchinger and Saxton (1989a); Jansson and Moon (2001)	Small-scale hydrothermal coupled model	Permafrost hydrothermal processes are finely delineated, but computationally complex, parameters are difficult to obtain, making it difficult to apply the model to large-scale catchment.

calculation speed and the depth of influence of permafrost freezing and thawing. Specifically, the soil surface layer was divided into 2 layers of 15 cm each in response to the relatively drastic hydrothermal changes; for the soil layer below the depth of 30 cm, it was divided into the 3rd–11th layers of 25 cm each. Through literature review and field inspection, it was comprehensively determined that soil evaporation occurs in layers 1–2 (0–30 cm), water absorption by the root systems of herbaceous plants and field crops covers layers 1–5 (0–105 cm), and water absorption by root systems of woody plants covers layers 1–11 (0–280 cm). Plant root density varied from layer to layer and decreased with depth. The thickness of each layer and the number of layers covered by roots can be adjusted according to actual conditions.

Furthermore, the calculation equation for soil ice content was introduced and then applied to the improvement of the Green-Ampt model, the Richards equation, and the heat transfer equation, deepening the mathematical description of permafrost hydrothermal processes from a physical mechanism perspective.

5.2.2.1 Improvement of the Infiltration and Runoff Processes in the Permafrost Layer

Based on rainfall intensity, the infiltration and runoff processes were divided into two types: Horton overland flow and saturation excess runoff. In the simulation process, based on the amount of precipitation, the runoff process was first divided into rainy and non-rainy periods. During the rainy period, vertical infiltration plays a dominant role in soil moisture movement, and the movement along the slope gradually becomes important after rainfall. Therefore, during the rainy period, the rainfall intensity was compared with the soil infiltration capacity. When the rainfall intensity exceeds the soil infiltration capacity, Horton overland flow occurs, calculated using a multi-layer Green-Ampt model. For the non-rainy period, in areas along riverbanks and low-lying areas where the soil is saturated or nearly saturated, an improved Richards equation was used for calculation.

In this study, the original calculation of the infiltration and runoff process was improved by considering the characteristics of permafrost freezing and thawing.

An 11-layer Green-Ampt model for frozen soil was established. During the process of permafrost freezing and thawing, the presence of ice in the soil changes the difference in soil saturated water content, $\Delta\theta_j$, thus the difference in soil water content can be written as equation (5.5):

$$\Delta\theta_j = \theta_{s,j} - \theta_{i,j} - \theta_{w0,j} \tag{5.5}$$

where $\theta_{s,j}$ is the soil saturation water content, $\theta_{i,j}$ is the soil ice content, $\theta_{w0,j}$ is the initial soil water content, cm³/cm³; the subscript j represents the soil layer number.

Similarly, the study referred to the SHAW model and used an effective hydraulic conductivity instead of the original saturated hydraulic conductivity.

$$K_{ef,j} = K_{s,j} \left(\frac{\theta_{s,j} - \theta_{i,j}}{\theta_{s,j}} \right)^{2b_j+3} \tag{5.6}$$

where $K_{ef,j}$ and $K_{s,j}$ represent the effective hydraulic conductivity and the saturated hydraulic conductivity, respectively, in cm/min; b_j is a parameter related to soil properties, obtained by fitting the soil moisture characteristic curve; the rest of the parameters are as before.

The permafrost moisture movement process (in integral form) was calculated with reference to equation (5.7):

$$\frac{\partial \theta_l}{\partial t} = \frac{\partial}{\partial z}\left[D(\theta_l)\frac{\partial \theta_l}{\partial z} - K(\theta_l) \right] - \frac{\rho_i}{\rho_w}\frac{\partial \theta_i}{\partial t} - S_r(z,t) \tag{5.7}$$

where θ_l represents the volumetric water content, cm³/cm³; θ_i represents the volumetric ice content in the soil, cm³/cm³; ρ_i and ρ_w, respectively, denote the density of ice and water, g/cm³; t is time, min; $K(\theta_l)$ is the soil hydraulic conductivity, cm/min; $D(\theta_l)$ is the moisture diffusivity, cm²/min; S_r is the source-sink term, min⁻¹; z is the vertical coordinate axis, positive downwards, cm.

5.2.2.2 Soil Hydrothermal Coupling Equation

Equation (5.8) was used to describe the relationship between soil water potential and temperature:

$$\psi(T) = \frac{L_f(T - T_f)}{gT} \tag{5.8}$$

The relationship between soil matrix potential and liquid water content was calculated using the Van Genuchten model (Van Genuchten, 1980), where the freezing point equation for permafrost was proposed and can be expressed as:

$$\theta_l = \theta_r + (\theta_s - \theta_r)\left\{ \frac{1}{1+\left[\dfrac{\alpha L_f(T - T_f)}{gT}\right]^n} \right\}^m \tag{5.9}$$

$$\theta_{i,j} = \theta_{t,j} - \theta_{l,j} \tag{5.10}$$

$$\theta_{l,j} = \min\left[\theta_{w0,j},\, \theta_r + (\theta_s - \theta_r)\left\{ \frac{1}{1+\left[\dfrac{\alpha L_f(T_s - T_f)}{gT_s}\right]^n} \right\}^m \right] \tag{5.11}$$

where $\Psi(T)$ represents the soil water potential, cm; L_f is the latent heat of fusion for ice, J/kg; T_f is the freezing point temperature, 273.15K; g is the acceleration due to gravity, m/s²; θ_r is the residual water content, θ_s is the saturated water content, θ_t is the total water content, θ_i is the solid water content, θ_{w0} is the liquid water content, θ_{w0} is the initial liquid water content, cm³/cm³; α is a parameter; T_s is the absolute soil temperature ($T_s = T + 273.15$), K; T is the soil temperature, °C; the subscript j refers to soil layers; other parameters are as previously defined.

5.2.2.3 Heat Transfer Equation

The soil heat transfer process was calculated as equation (5.12):

$$c_v \frac{\partial T}{\partial t} = \frac{\partial}{\partial z}\left[\lambda \frac{\partial T}{\partial z}\right] + L_i \rho_i \frac{\partial \theta_i}{\partial t} \tag{5.12}$$

where L_i is the latent heat of fusion for ice, J/g; T is the soil temperature, °C; z is the vertical coordinate axis, cm; c_v is the specific heat capacity of the soil, J/(cm³/°C); t is time, min; λ is the thermal conductivity of unsaturated soil, W/(cm/K). Other parameters are as previously defined. The soil heat conduction equation is discretized using the finite difference method, with the lower boundary being a fixed temperature boundary and the upper boundary being a heat flux boundary.

The unsaturated thermal conductivity λ was calculated by equations (5.13) and (5.14):

$$\lambda = (\lambda_{sat} - \lambda_{dry})K_e + \lambda_{dry} \tag{5.13}$$

$$\lambda_{dry} = \frac{0.135\rho_b + 64.7}{2700 - 0.947\rho_b} \tag{5.14}$$

Saturated soil thermal conductivity was calculated separately for both frozen and unfrozen soil types. The calculation formula for unfrozen soil was equation (5.15):

$$\lambda_{sat} = \lambda_m^{1-\phi} \lambda_w^{\phi} \tag{5.15}$$

When the soil is frozen, the calculation formula was equation (5.16):

$$\lambda_{sat} = \lambda_m^{1-\phi} \lambda_w^{wv} \lambda_i^{\phi-wv} \tag{5.16}$$

where λ_{sat} is the saturated soil thermal conductivity, W/(cm/K); λ_{dry} is the dry soil thermal conductivity, W/(cm/K); ϕ is the porosity, cm³/cm³; wv is the volume of liquid water, cm³/cm³; λ_m, λ_w, and λ_i are the thermal conductivities of minerals, liquid water, and solid water, respectively, W/(cm/K); K_e is the Kersten number. In addition, the soil-specific heat capacity, latent heat of dissolution, thermal conductivity, porosity, and particle composition required in the model can be measured and obtained.

For the glacier-covered area, the effect of the glacial heat transfer process on soil temperature was ignored, i.e., the heat flux was zero, because the glacier is thicker and the air temperature has less effect on soil temperature. The model's energy balance used daily data, and the equilibrium method was used for the calculation of sensible heat. It first calculated the net radiation, ground heat flux, latent heat flux, and the melting heat flux of glacier and snow, and finally, the sensible heat flux was solved using the energy balance equation.

5.2.2.4 Subsurface Runoff Calculation

In the original WEP-L model, shallow groundwater was numerically calculated in two dimensions using the Boussinesq equation, with recharge from the unsaturated soil layer, groundwater withdrawals, and groundwater outflows (or recharge from the river) as source terms. Considering that a decrease in ground temperature significantly affects the groundwater movement, the improved model assumes that the groundwater hydraulic conductivity K_g, exponentially decreases with a decrease in ground temperature once T_s falls below a certain critical temperature T_c (below zero). The calculation formula is equation (5.17):

$$K_g = K_{g0} \exp[a(T_s - T_c)] \tag{5.17}$$

where K_{g0} is the hydraulic conductivity of groundwater under normal climatic conditions ($T_s \geq T_c$); a is a parameter. T_c and a are determined based on the calibration of observed winter runoff data.

5.3 VALIDATION OF THE DISTRIBUTED HYDROTHERMAL COUPLING MODEL

5.3.1 Study Area

The SAYR covers the area above the ZhiMenDa (ZMD) hydrological station, located at 90°E to 98°E and 31°N to 36°N (Figure 5.1). The terrain slopes downward from west to

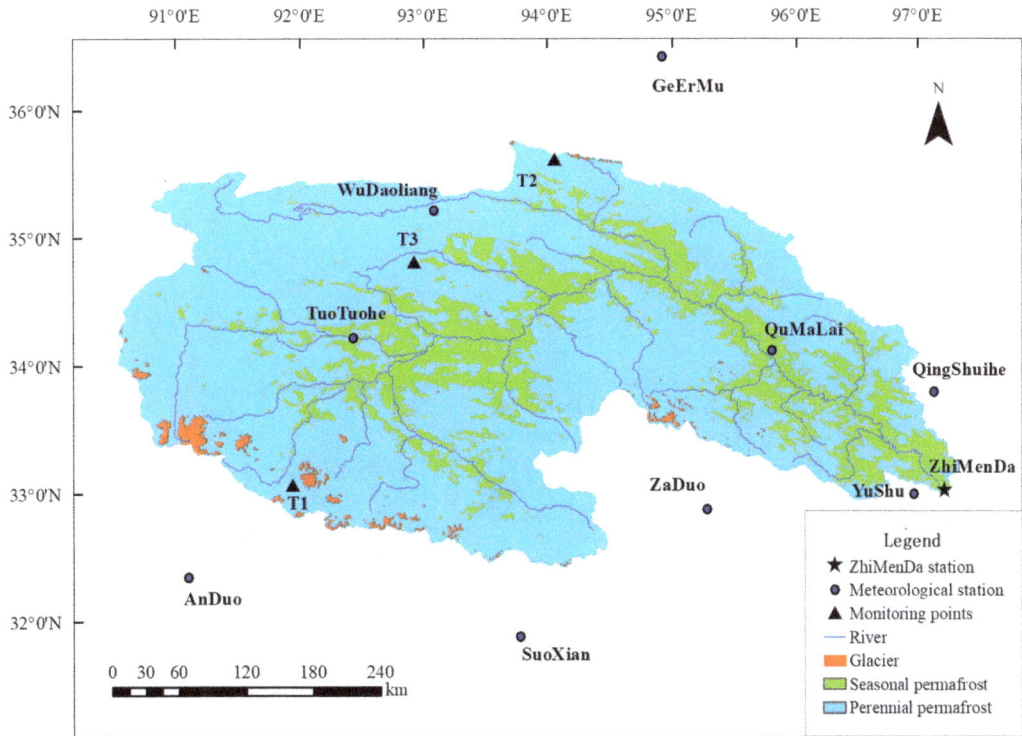

FIGURE 5.1   Schematic of the SAYR.

east, with an average elevation of about 4700 m. It experiences strong solar radiation and has an average annual precipitation of about 370 mm, with the main soil types being alpine grassland and alpine meadow. The area of perennial and seasonal permafrost zones is 108,000 km^2 and 31,000 km^2, accounting for 78% and 22% of the total area, respectively. To verify the accuracy of the WEP-SAYR model in simulating soil water and heat transfer processes, comprehensive permafrost monitoring data provided by Zhao et al. (2021) from the National Tibetan Plateau Data Center (Hu et al., 2017) are collected, selecting three monitoring sites T1 (33.07°, 91.94°), T2 (35.62°, 94.06°), and T3 (34.82°, 92.92°). These sites are located in perennial permafrost areas, with monitoring elements including daily scale air temperature, precipitation, soil temperature, and humidity, covering the period from December 2012 to December 2014. Monitoring depths vary among the sites, with T1 at 50cm and 140 cm; T2 at 10 cm, 40 cm, and 100 cm; and T3 at 10 cm, 50 cm, and 90 cm depths. It should be noted that the 90 cm depth at site T3 is only monitored for soil temperature, not soil moisture content.

5.3.2 Collection of Basic Information

The main data required for constructing distributed hydrological models include hydrometeorological data, land use/cover data, topographic and geomorphologic data, soil type data, glacier and snow data, and river channel information data, as seen in Table 5.2. All parameters in the model, including leaf area index, NDVI, and land use

TABLE 5.2   Basic Information Required for Constructing the WEP-SAYR Model

Data Type	Item	Data Content
Topography and Geomorphology	DEM	1 km × 1 km spatial resolution
River Network	Water system	Real water system information
	River Channel Cross-section	Typical river channel cross-section information
Meteorological Hydrology (1980–2020)	Rainfall	Daily data at meteorological stations, 1980–2020
	Wind speed	
	Air temperature	
	Sunshine	
	Humidity	
	Monthly actual flow measurement	1980–2020
	Daily actual flow measurement	2006–2020
Land Use/Cover data (1980, 1990, 2005, 2010, 2015, 2020)	Land use	1:100,000 land use map
	Vegetation index	Decadal vegetation index
	Vegetation cover	Decadal vegetation cover
	Leaf Area Index	Decadal leaf area index
Glacier	Glacier area, range, ice storage	China's First Glacier Inventory
Snow	Snow depth simulated starting in 1980	Remote sensing product of Sanjiangyuan snow depth (1980–2020)
Soil Type	Soil type map	National Second Soil Survey information map, resolution 1:100000
Hydrogeological Information	Hydrogeological Distribution Map of China	Hydrogeological parameters, lithology distribution, and aquifer thickness

types, are derived from remote sensing interpretation data. The soil types in the model are divided into sand, loam, and clay, with specified soil hydrothermal characteristic parameters.

5.3.3 Division of Calculation Units

The SAYR was divided into 1464 sub-basins. Each sub-basin was further divided into 1–5 contour bands based on elevation changes, totaling 4937 contour bands in the whole catchment (Figure 5.2). These bands were identified as the basic computation units for the model, with an average area of about 27.95 km². During the modeling process, to account for the differences in hydrothermal processes between permafrost and seasonally frozen ground, the lower boundary value in the soil heat transfer equation was set to 0°C at the

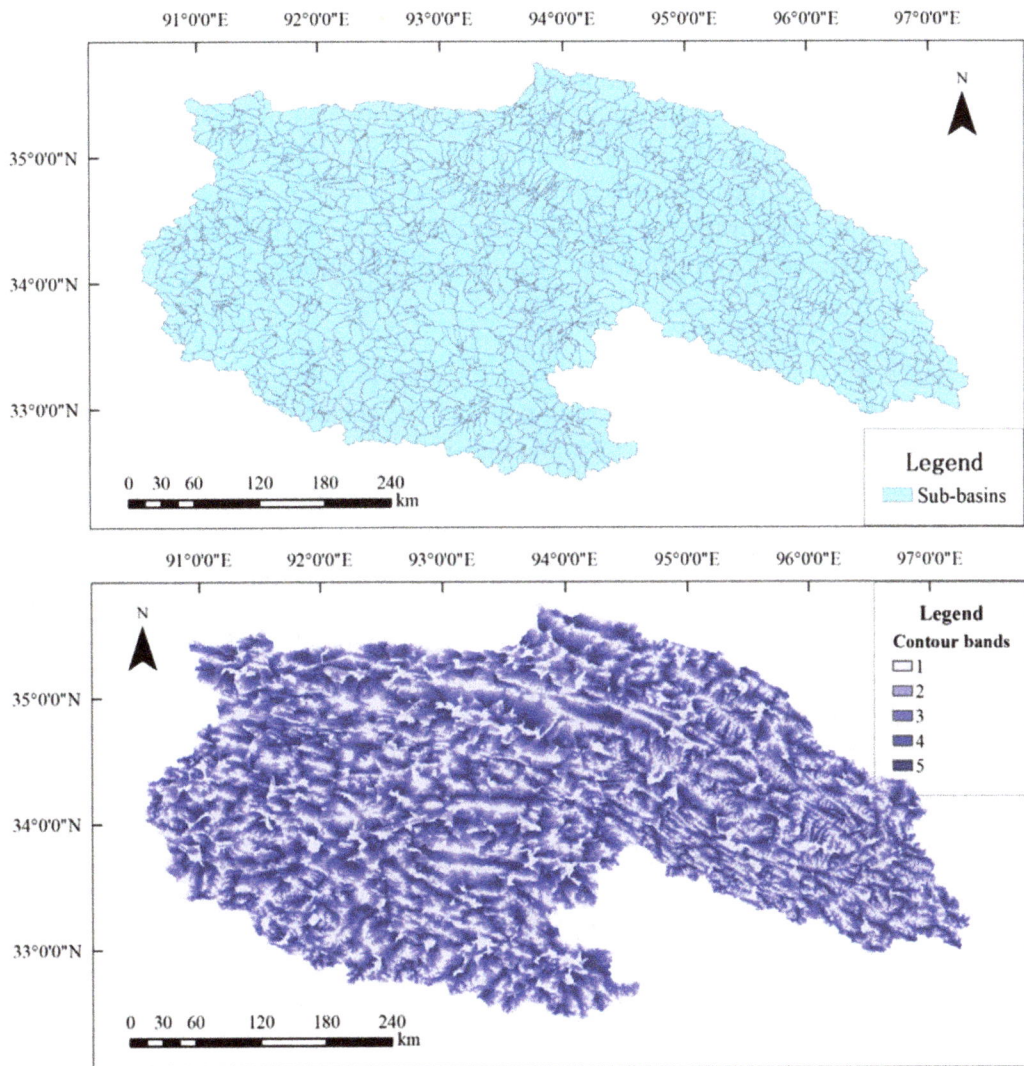

FIGURE 5.2   Division of calculation units in the SAYR.

junctions between the two areas, with perennial permafrost zones having negative temperatures and seasonal permafrost zones having lower boundary temperatures above 0°C. The lower boundary temperature varies with altitude. Therefore, the lower boundary temperature of the soil in both calculation units was corrected for altitude.

5.3.4 Parameter Calibration and Validation

The parameters of the WEP-SAYR model mainly included four categories: ground surface and river system parameters such as Manning coefficient and riverbed permeability coefficients; vegetation parameters such as Leaf Area Index, stomatal resistance; soil water movement parameters such as soil saturated hydraulic conductivity, soil porosity, aquifer thickness, field capacity, aquifer conductivity; soil thermal conductivity. Due to the spatial variability of parameters, the model used unit averages or effective parameters, thus, some key parameters need to be adjusted through validation.

The runoff process modeling selected actual monthly and daily data from the ZMD station. For the monthly runoff, the calibration period was 1980–2000, and the validation period was 2001–2020. The daily runoff data from 2006 to 2020 were also validated. The WEP-SAYR model employed a variable time step approach, using 1 hour for infiltration processes with rainfall intensity exceeding 5 mm, 6 hours for hill slope and river channel convergence, and 1 day for the rest. The Nash-Sutcliffe efficiency coefficient (NSE) and the relative error of runoff (RE) were used to assess model performance.

5.3.5 Analysis of Simulation Results

5.3.5.1 Accuracy of Spatial Interpolation of Station Meteorological Data

The meteorological data from monitoring point T1 (no meteorological data for T2 and T3), including temperature and precipitation, were utilized to validate the results of spatial interpolation of the national station meteorological data, as seen in Table 5.3. It can be seen that the three-dimensional spatial interpolation results of temperature, taking into account the impact of altitude, are close to the measured values, with a RE value of 2.55% and an NSE value of 0.96. The error for precipitation was larger, with a RE value of 24.29% and an NSE value of 0.53. Compared to temperature, precipitation has stronger spatial heterogeneity, and the scarcity of national meteorological stations within the catchment leads to larger spatial errors in precipitation data.

5.3.5.2 Validation of Soil Temperature Simulation Results

The original WEP-L model fails to consider the process of hydrothermal transfer processes in permafrost regions, consequently, it is incapable of accurately simulating temperature

TABLE 5.3 Results of Temperature and Precipitation Distribution Validation

Data Point	Meteorological Element	Evaluation Indicators	
		RE (%)	NSE
T1	Temperature	2.55	0.96
	Precipitation	24.29	0.53

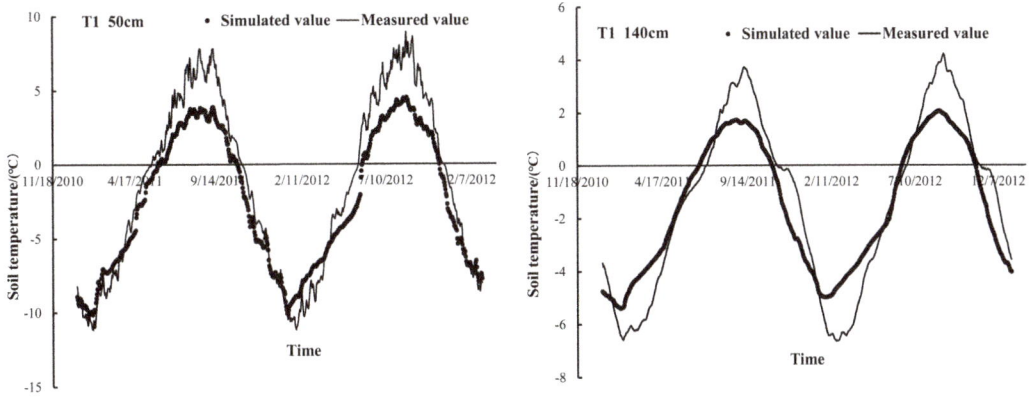

FIGURE 5.3 Soil temperature monitoring and simulation results over time (T1, T2, and T3 represent different monitoring points, respectively).

variations at different soil depths. The WEP-SAYR model's simulation of soil temperatures at various depths compared to monitoring results is shown in Figure 5.3 and Table 5.4. It is important to note that, to meet the requirements of catchment-scale simulation, the input data, such as air temperature and precipitation, were used as the result of spatial interpolation from national meteorological data. The data represents the face-averaged value of the monitoring point in the calculation unit, not the measured value at the monitoring points.

It can be seen that the improved WEP-SAYR can still reflect the soil temperature change process well, although the simulation results have some errors with the point observations due to the inconsistency of the input data scale. The *RE* values between the simulated and observed temperatures of different soil layers ranged from −62.01% to 49.14%, and the values of *NSE* were all above 0.72, indicating that the model's simulation results at the contour bands scale can essentially depict the area's soil temperature and its variation with depth. The changes of soil temperature with air temperature at different depths were consistent, in which the surface soil responded faster to the changes of air temperature, and the deeper the soil, the slower the response.

TABLE 5.4   Results of Soil Temperature and Moisture Simulation Validation

Monitoring Point	Soil Depth (cm)	Soil Moisture Content		Soil Temperature	
		RE(%)	*NSE*(−)	*RE*(%)	*NSE*(−)
T1	50	−16.23	0.41	−1.18	0.90
	140	18.69	0.40	37.31	0.86
T2	10	−1.01	0.44	−16.10	0.94
	40	−3.26	0.43	49.14	0.89
	100	34.00	0.42	−62.01	0.76
T3	10	−29.85	0.41	12.01	0.67
	50	−17.66	0.64	0.42	0.82
	90	—	—	0.41	0.72

5.3.5.3 Validation of Soil Moisture Simulation Results

The soil moisture simulation results of WEP-SAYR were compared to monitoring data, as shown in Figure 5.4 and Table 5.4. It is important to note that the values identified as missing or erroneous by the data provider were excluded from the analysis presented by Hu et al. (2017). Considering the soil freezing and thawing process, the *RE* values between the

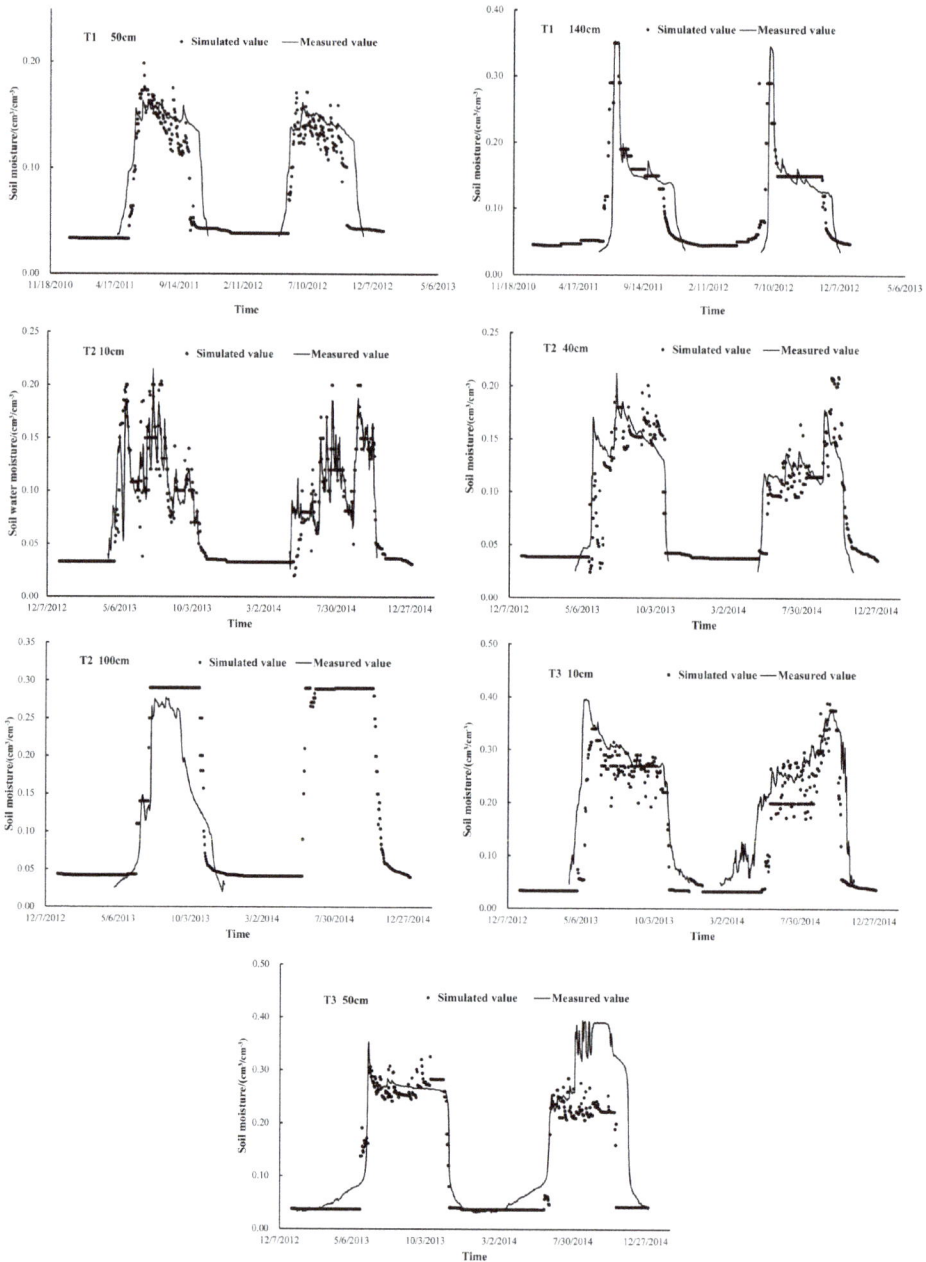

FIGURE 5.4   Monitoring and simulation results of soil water content change over time (T1, T2, and T3 represent different monitoring points, respectively).

simulated and observed soil moisture at different depths ranged from −29.85% to 34.00%, with *NSE* values exceeding 0.4. Given the inconsistency in the scale of input data, the improved model essentially depicted the moisture change trends in soil layers at different depths. Compared to *NSE* values for soil temperature simulation, the *NSE* values for soil moisture simulation were lower. This discrepancy may be due to soil moisture changes being influenced not only by soil temperature but also by precipitation. The inconsistency between the model's input values for precipitation and the actual monitoring values at specific points caused more disturbance to the soil moisture simulation.

5.3.5.4 Validation of monthly/daily Runoff Simulation

The WEP-L model and the WEP-SAYR model were used to simulate monthly and daily runoff at ZMD station from 1980 to 2020, validated by actual daily/monthly measured data, as listed in Table 5.5.

From the results, the simulation effect of the WEP-SAYR model on the monthly and daily runoff was significantly improved by considering the processes of the glacial and snowmelt runoff and soil freezing and thawing. The *RE* values for monthly runoff during the calibration and validation periods were 0.2% and −6.8%, respectively, with the *NSE* values of 0.84 and 0.83; the *NSE* value for daily runoff was 0.80, with *RE* within ±10%. In contrast, the simulation results of monthly/daily runoff processes by the WEP-L model were significantly worse, with *RE* values ranging from −5.4% to −15.5% and *NSE* values below 0.7. The improvements in the model significantly enhanced the accuracy of simulating base flow and peak flow rates in rivers.

5.3.5.5 Validation of Monthly Evaporation Simulation

The Global Land Evaporation Amsterdam Model (GLEAM) data is a commonly used global terrestrial evapotranspiration product (https://www.gleam.eu/). This study utilized GLEAM data for the inverse modeling of monthly evapotranspiration from 1980 to 2020, to validate simulated values of WEP-SAYR, as shown in Table 5.6 and Figure 5.5. Throughout

TABLE 5.5   Results of Flow Calibration and Validation at the ZMD Station

		WEP-L Model		WEP-SAYR Model	
Runoff	**Period**	**RE(%)**	**NSE**	**RE(%)**	**NSE**
Monthly Data	1980–2000	−5.4	0.73	0.2	0.84
	2001–2020	−13.7	0.71	−6.8	0.83
	1980–2020	−10.1	0.71	−6.1	0.84
Daily Data	2006–2020	−14.6	0.65	−6.2	0.80

TABLE 5.6   Results of Monthly Evapotranspiration Calibration and Validation in the SAYR from 1980 to 2020

		Inversion of GLEAM Data	
Evapotranspiration	**Period**	**RE(%)**	**NSE**
Monthly Average	Calibration Period (1980–2000)	8.7	0.83
	Validation Period (2001–2020)	10.8	0.77
	Simulation Period (1980–2020)	9.7	0.80

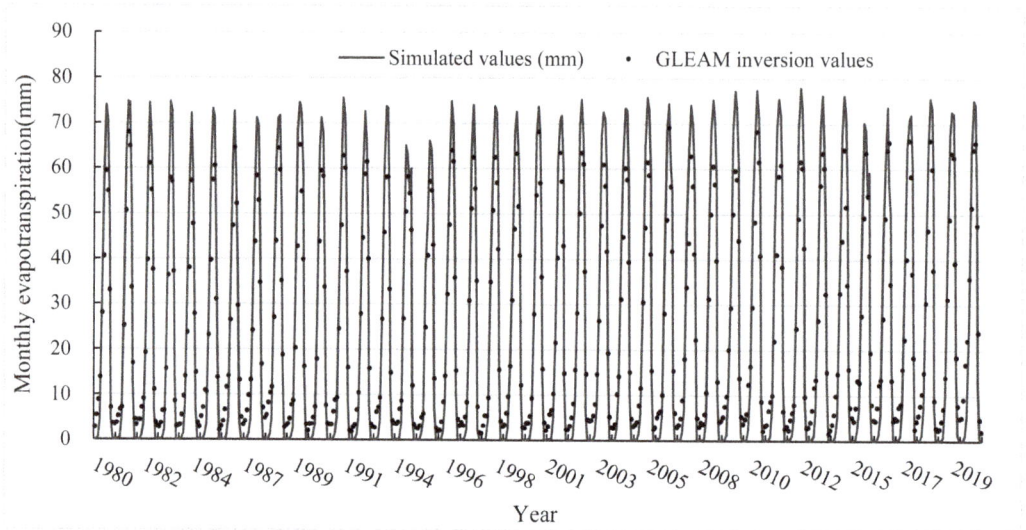

FIGURE 5.5   Validation of monthly evapotranspiration in the SAYR from 1980 to 2020.

the entire simulation period, the value of *RE* between the GLEAM inversion values and the simulated values was 9.7%, with a *NSE* value of 0.80. These results indicated that the simulation results can accurately reflect the seasonal variation pattern of evapotranspiration.

5.4 ANALYSIS AND DISCUSSION OF RESULTS

Against the backdrop of global warming, the hydrological processes of the cryosphere, dominated by glaciers, snow, and permafrost, undergo dramatic changes, thereby affecting runoff and water resources. The study constructed the WEP-SAYR model by integrating glacial and snow meltwater and permafrost hydrothermal transfer processes into the traditional distributed hydrologic model WEP-L. For the simulation of permafrost hydrothermal processes, the models represented by the CoLM scheme calculated the relationship between soil liquid water content and water potential without considering residual water content. This can lead to an underestimation of the minimum liquid water content at very low temperatures, even resulting in zero values, which is not realistic. This study adopted the van Genuchten model and proposed a freezing point equation, which in turn allows the permafrost hydrothermal equations of motion to be solved jointly and improves the shortcomings of previous models when calculating the minimum water content. Meanwhile, regarding models like SHAW and CoLM, the soil hydrothermal transfer process is calculated using the mass conservation equations and energy conduction model, and a model for calculating hydrothermal characteristic parameters was given. On the other hand, the degree-day factor method for simulating glacier melting processes serves as an effective quantitative tool when there is insufficient acquisition of hydrothermal parameters for glaciers and snow. Moreover, since the WEP-SAYR model divides the catchment into many contour belts, it fully reflects the heterogeneity of glacier degree-day factors with changes in elevation and latitude, with results consistent with research findings by Qiao et al. (2010). Considering the model's simulation analysis process is physically meaningful, it is beneficial for its application and extension to other basins within China's cryosphere.

In previous studies, hydrological models have been applied in the Yangtze River source region, but the consideration of cryospheric elements is relatively simplistic. Yang et al. (2022) applied the SWAT model to the Yangtze River source region, but the model did not consider glaciers, snow, and permafrost. Although the model simulates the runoff process well, it cannot depict the dynamic changes in permafrost thermohydraulic coupling or the melting processes of glaciers/snow. Wang et al. (2023) applied the GBEHM model across the entire Tibetan Plateau, but the model simplifies the depiction of permafrost thermo-hydraulic coupling relationships and cannot simulate the coupled changes between permafrost moisture and temperature. Compared to previously applied models, WEP-SAYR illustrates the impact of glaciers, snow, and permafrost on hydrological processes in detail, capable of simulating glacier/snow melting, dynamic changes in permafrost water and heat coupling, and runoff dynamics, with physically meaningful and easily accessible model parameters.

However, there are still parts of the model that need to be improved or proven. First, the groundwater storage in the SAYR is in various forms, including water above the permafrost layer, water between the permafrost layers, and water below the permafrost layer, resulting in a complex mechanism of water-heat transfer. Limited by the lack of data, this model did not conduct an analysis and simulation of the groundwater hydrothermal transfer mechanisms, only crudely describing the groundwater outflow process through a simple functional relationship between groundwater conductivity and air temperature, and optimizing parameters through baseflow simulation. Second, the processes of glacier and snow motion are complex, involving windblown snow, sublimation, and compaction into ice. This study still adopts the simple temperature index model to depict the melting process of glaciers and snow. Regarding the impact of snow cover and snowmelt infiltration on glacier melting, the simulation currently follows the sequence where glacier melting begins only after snowmelt, without considering the interaction between them, increasing the uncertainty in basin-scale cryosphere simulations.

REFERENCES

Dai, Y., Zeng, X., Dickinson, R., et al., 2003. The common land model. Bulletin of the American Meteorological Society, 84(8), 1013–1024.

Ding, B., Yang, K., Yang, W., et al., 2017. Development of a water and enthalpy budget-based glacier mass balance model (WEB-GM) and its preliminary validation. Water resources research, 53(4), 3146–3178.

Ding, Y., Zhao, Q., Wu, J., et al., 2020. The future changes of Chinese cryospheric hydrology and their impacts on water security in arid areas. Journal of glaciology and geocryology, 42(1), 23–32.

Flerchinger, G.N. and Saxton, K.E., 1989a. Simultaneous heat and water model of freezing snow-residue-soil system. I. Theory and development. Trans ASAE, 32, 565–571.

Fontaine, T.A., Cruickshank, T.S., Arnold, J.G., et al., 2002. Development of a snowfall–snowmelt routine for mountainous terrain for the soil water assessment tool (SWAT). Journal of hydrology, 262(1–4), 209–223.

Gao, B., Yang, D., Qin, Y., et al., 2018. Change in frozen soils and its effect on regional hydrology, upper Heihe basin, northeastern Qinghai–Tibetan Plateau. The cryosphere, 12(2), 657–673.

Hu, G., Zhao, L. and Wu, X. et al., 2017. Comparison of the thermal conductivity parameterizations for a freeze-thaw algorithm with a multi-layered soil in permafrost regions. Catena, 156, 244–251.

Jansson, P.E. and Moon, D.S., 2001. A coupled model of water, heat and mass transfer using object orientation to improve flexibility and functionality. Environmental modelling and software, 16(1), 37–46.

Jia, Y., Wang, H., Zhou, Z., et al., 2006. Development of the WEP-L distributed hydrological model and dynamic assessment of water resources in the Yellow River basin. Journal of hydrology, 331(3–4), 606–629.

Liu, Q., Bai, S., Huang, J., et al., 2004. A thermodynamic coupling scheme for ice-ocean model. Acta oceanologica sinica, 26(6), 13–21.

Li, J., Zhou, Z., Wang, H., et al., 2019. Development of WEP-COR model to simulate land surface water and energy budgets in a cold region. Hydrology research, 50(1), 99–116.

Liu, H., Hu, Z., Yang Y., et al., 2015. Simulation of the freezing-thawing processes at Nagqu Area over Qinghai-Xizang Plateau. Plateau meteorology, 34(3), 676–683.

Liu, Y., Xu, J., Lu, X., et al., 2020. Assessment of glacier- and snowmelt-driven streamflow in the arid middle Tianshan Mountains of China. Hydrological processes, 34(12), 2750–2762.

Ma, H., 2003. A test of snowmelt runoff model (SRM) for the Gongnaisi River basin in the western Tianshan Mountainss, China. Chinese Science bulletin, 48(20), 2253.

Qiao, C., He, X. and Ye, B., 2010. Study of the degree-day factors for snow and ice on the Dongkemadi Glacier, Tanggula Range. Journal of glaciology and geocryology, 32(2), 257–264.

Terink, W., Lutz, A.F. and Simons, G.W.H. et al., 2015. SPHY v2.0: Spatial processes in Hydrology. Geoscientific model development, 8(7), 2009–2034.

Van Genuchten, M.T., 1980. A closed-form equation for predicting the hydraulic conductivity of unsaturated soils. Soil science society of America journal, 44(5), 892–898.

Wang, T., Yang, D., Yang, Y., et al., 2023. Unsustainable water supply from thawing permafrost on the Tibetan Plateau in a changing climate. Science bulletin, 68(11), 1105–1108.

Yang, Y., Ma, L., Li, S., et al., 2022. Construction of SWAT model database in the Yangtze River source region and calibration and verification of the model. Journal of Anhui University (Natural science edition), 46(4), 76–84.

Zhang, Y., Liu, S. and Wang, X., 2019a. A dataset of spatial distribution of degree-day factors for glacier in High Mountain Asia. China scientific data, 4(3), 141–151.

Zhao, L., Zou, D., Hu, G., et al., 2021. A synthesis dataset of permafrost thermal state for the Qinghai–Tibet (Xizang) Plateau, China. Earth system science data, 13(8), 4207–4218.

Zhou, G., Cui, M., Wan, J., et al., 2021a. A review on snowmelt models: Progress and Prospect. Sustainability, 13(20), 11485.

Zhou, Z., Liu, Y., Li.,Y., et al., 2021b. Distributed hydrological model of the Qinghai Tibet Plateau based on the Hydrothermal coupling: I: Hydrothermal coupling simulation of "snow-soil-sand gravel layer" continuum. Advances in water science, 32(1), 20–32 (in Chinese).

CHAPTER 6

Development and Validation of China Water Assessment Model

6.1 DETERMINATION OF BASIC DATA AND CALCULATION UNITS

6.1.1 Data Collection and Description

The basic data to be prepared for driving the China Water Assessment Model (CWAM) included hydro-meteorology, topography information, river network system, soil information, land use, hydrogeology, vegetation, etc., as listed in Table 6.1.

6.1.1.1 Hydro-meteorology

The meteorological data of 2000 national meteorological stations (1956–2017) with five items of rain/snow, air temperature, sunshine hours, vapor pressure/relative humidity, and wind speed daily were provided by the National Meteorological Information Center of China. The integrity of the collected daily data was checked. If there was missing data at a station, the nearest station was used for interpolation and extrapolation. Using the observed sunshine hours, the short-wave and long-wave radiations were estimated (Maidment, 1992). Moreover, the initial hydrological data, including initial river flow, soil moisture content, and groundwater table, were estimated by consulting the Annual Hydrological Report of China.

6.1.1.2 Topography Information

The Shuttle Radar Topography Mission (SRTM) with a spatial resolution of 90 m was obtained from the Geospatial Data Cloud site, Computer Network Information Center, Chinese Academy of Sciences (http://www.gscloud.cn). Then, the 90 m × 90 m grid was resampled to 1 km × km DEM raster.

DOI: 10.1201/9781003646648-6

TABLE 6.1   Basic Data of the CWAM

Data Types and Elements		Data Sources	Data Periods
Hydro-meteorology	Rain/snow	Daily monitoring data from meteorological stations	1956–2017
	Wind speed		
	Air temperature		
	Sunshine hours		
	Relative humidity		
	Initial river flow	Annual Hydrological Report of China; Model simulation results	2000
	Initial soil moisture content		—
	Initial groundwater table		2010–2017
Topography Information		90 × 90m DEM	2000
River Network System	River network distribution	National Geography Database	2011
	Cross-section information	Annual Hydrological Report of China	2017
Soil Information	Soil types	National Second Soil Survey Data	2004
Land Use	Land use types	Chinese Academy of Sciences	1980, 1990, 1995, 2000, 2005, 2010, and 2015
Hydrogeology	Hydrogeology unit division, permeability and storage coefficient	Distribution Map of Hydrogeology in China	1980s
Vegetation	Fractional vegetation cover	MODIS data	2001–2017
	Leaf area index		2001–2017

6.1.1.3 River Network System

The surveyed river network was obtained from the National Geography Database with a scale of 1:250,000. Besides, the representative cross-section information of the main rivers was obtained from the Annual Hydrological Report of China. All cross-sections of rivers were generalized to an isosceles trapezoid, and the upper and lower bottom widths and maximum depths of the cross-sections were estimated based on measured information.

6.1.1.4 Soil Information

The map of 1km grid soil types, along with their characteristic parameters, was obtained from the National Second Soil Survey Data and Soil Types of China. In accordance with international standards, the country's soils were classified into four major categories and twelve types: sandy soils: sandy or loamy sandy soils; loamy soils: sandy loam, loamy loam, and chalky loam; clay loam: sandy clay loam, clay loam, and chalky clay loam; and clayey soils: sandy clayey soils, loamy clays, chalky clays, clayey clays, and heavy clayey clays.

6.1.1.5 Land Use

The national land-use data of seven periods (1980, 1990, 1995, 2000, 2005, 2010, and 2015) with 1km spatial resolution were obtained from the Chinese Academy of Sciences (http://www.dsac.cn/DataProduct). The data showed that the proportion of grassland is the

TABLE 6.2   Changes in the Proportion of Land Area in the Six Types of Land Use in China

Area Proportion	1980	1990	1995	2000	2005	2010	2015
Crop	18.5%	18.6%	18.5%	19.0%	18.9%	18.8%	18.8%
Forest	23.8%	23.7%	24.0%	23.6%	23.6%	23.6%	23.6%
Grass	32.1%	32.0%	31.5%	31.7%	31.6%	31.6%	31.5%
Water Body	3.0%	2.8%	2.8%	2.9%	2.9%	2.9%	2.9%
Residential and Industrial Building Land	1.6%	1.6%	1.8%	1.8%	2.0%	2.1%	2.3%
Unutilized Land	21.0%	21.1%	21.5%	21.1%	21.0%	21.0%	20.9%

highest in the country and has decreased over the years, but the change is small. It was followed by woodland and unused land, which showed a slight decreasing trend. Arable land and waters have fluctuated slightly, and the proportion of residential and industrial land was the smallest, but there has been a sustained and significant increase over the years. Besides, there are significant differences in different land use compositions and changes in 10 Class I WRRs, as shown in Table 6.2.

6.1.1.6 Hydrogeology
The data included the hydrogeology unit division, permeability, and storage coefficient. They were obtained from the Distribution Map of Hydrogeology in China and the second national water resources survey.

6.1.1.7 Vegetation
The monthly values of LAI and FVC from 2001 to 2017 were obtained using Moderate Resolution Imaging Spectroradiometer (MODIS) satellite data (http://www.dsac.cn/DataProduct).

6.1.2 Determination of Computation Units
Using the sub-basin division method for large-scale distributed hydrology modeling, proposed in Chapter 2, China was divided into 21,768 sub-basins. Furthermore, the river network in some desert areas was fused to finalize the number of computational units in the CWAM as 19,406. The average area of sub-basins is about 490 km², and the maximum and minimum areas are 5512 and 4 km², respectively.

According to the extent of the plains and mountainous areas of the country, these sub-basins were further divided into several contour belts. For a sub-basin located in the mountainous area, the number of contour belts will be a maximum of six and a minimum of one, keeping the contour belts as even as possible in size. Due to the less undulating topography, sub-basins in the plain area are no longer done as elevation subdivisions, and each sub-basin is equivalent to a contour belt. As a result, the 19,406 sub-basins of the CWAM were further divided into 81,687 contour belts, which were used as computation units for the model, as shown in Figures 6.1 and 6.2.

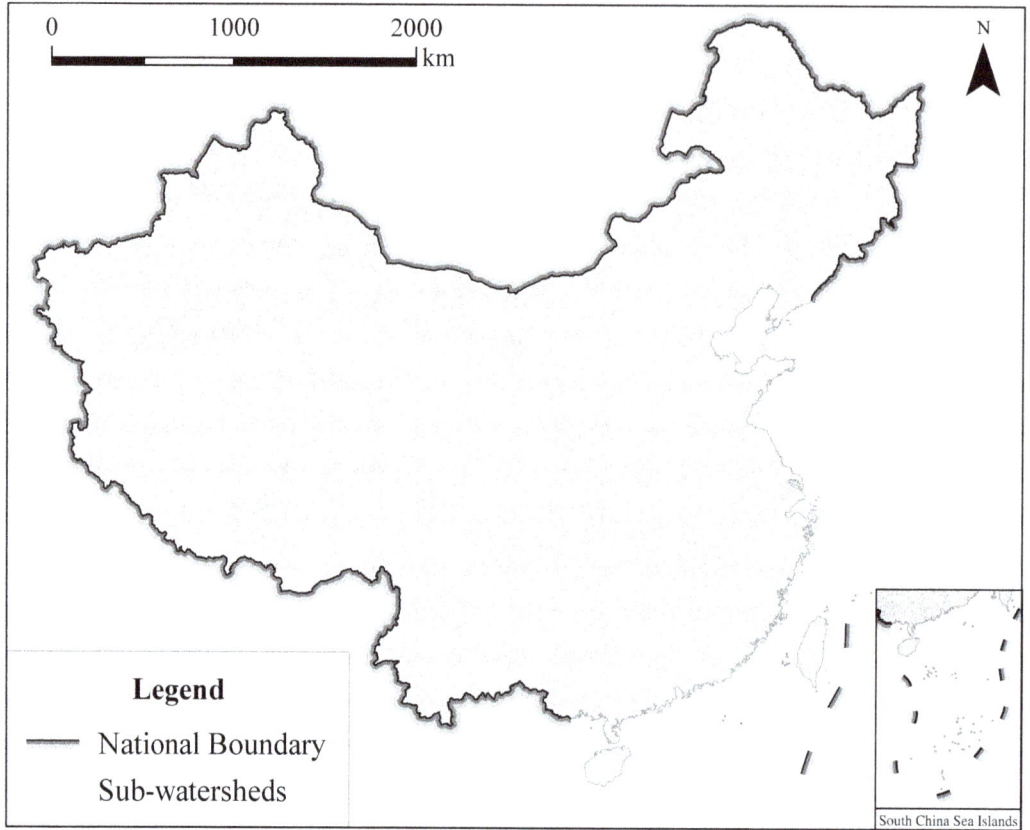

FIGURE 6.1   Sub-basins division in China.

6.1.3 Spatial Three-Dimensional Interpolation of Meteorological Data

Using the reverse distance square (RDS) method that considers the elevation effect to get the precipitation and air temperature in each sub-basin. The method is formulated as equations (6.1)–(6.3).

$$D = \sum_{i=1}^{m} \lambda_i \left(D_i + \frac{dM}{dz}(z_0 - z_i) \right) \tag{6.1}$$

$$\lambda_i = \frac{d_i^{-2}}{\sum_{i=1}^{m} d_i^{-2}} \tag{6.2}$$

$$d_i = \sqrt{(x_i - x_0)^2 + (y_i - y_0)^2} \tag{6.3}$$

where dM/dz is the linear change rate of precipitation and air temperature with elevation, z is the elevation, subscript 0 denotes the sub-basin to be interpolated, i is the reference

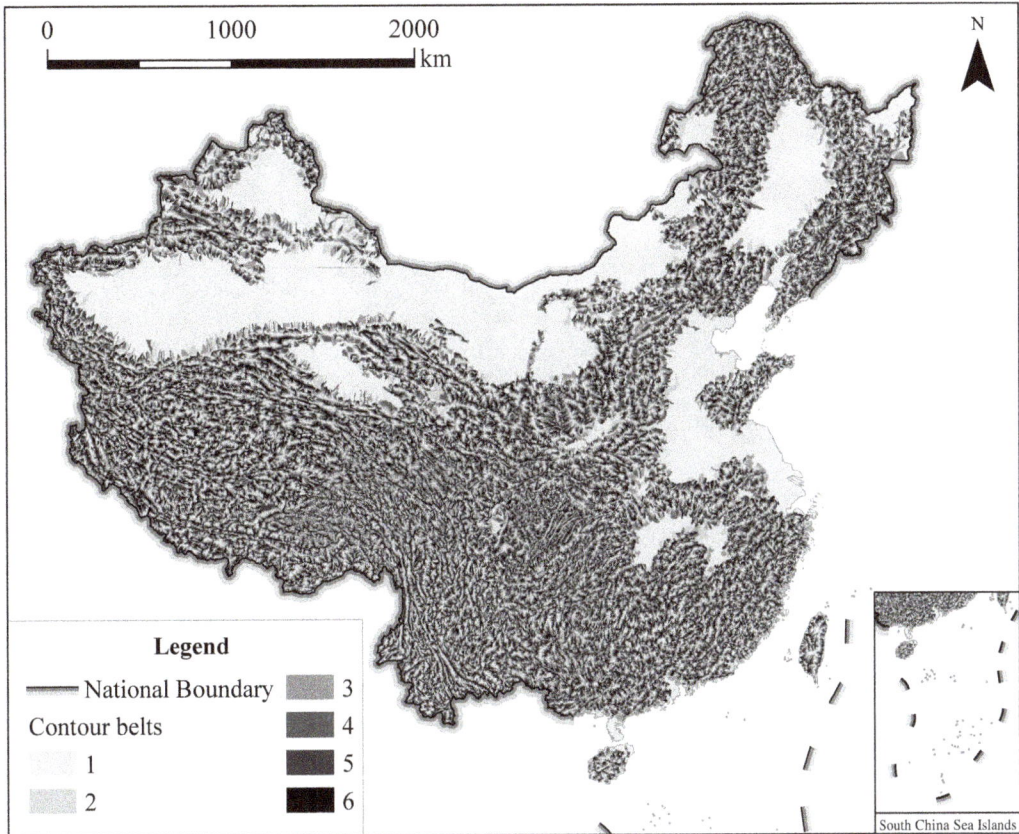

FIGURE 6.2   Contour belts division in China.

station, D is the meteorological data, λ_i is the weight of the ith reference station, and d_i is the plane projection distance between the ith reference station and the interpolation point.

In this chapter, the linear rate of change of precipitation and temperature with elevation was determined for each of 80 Class II WRRs. First, for each Class II WRR, average annual precipitation and temperature values and corresponding elevations were collected for all sub-basins located in this region. Second, the linear correlation between mean annual precipitation and air temperature, and elevation was assessed. If the correlation was significant (i.e., p-value < 0.05), the dM/dz values were calculated using equations (6.1)–(6.3). Otherwise, the dM/dz values were considered to be 0. Figures 6.3 and 6.4 illustrate the dM/dz values of precipitation and air temperature in each of 80 Class II WRRs, respectively.

For other meteorological data, such as the daily wind speed, sunshine hours, and humidity, the RDS method was simply adopted, and the elevation impact was ignored. Moreover, for several regions with few meteorological stations, like western China (including the Tibetan Plateau), the Global Precipitation Mission (GPM) data and station observation precipitation were used to comprehensively determine local precipitation.

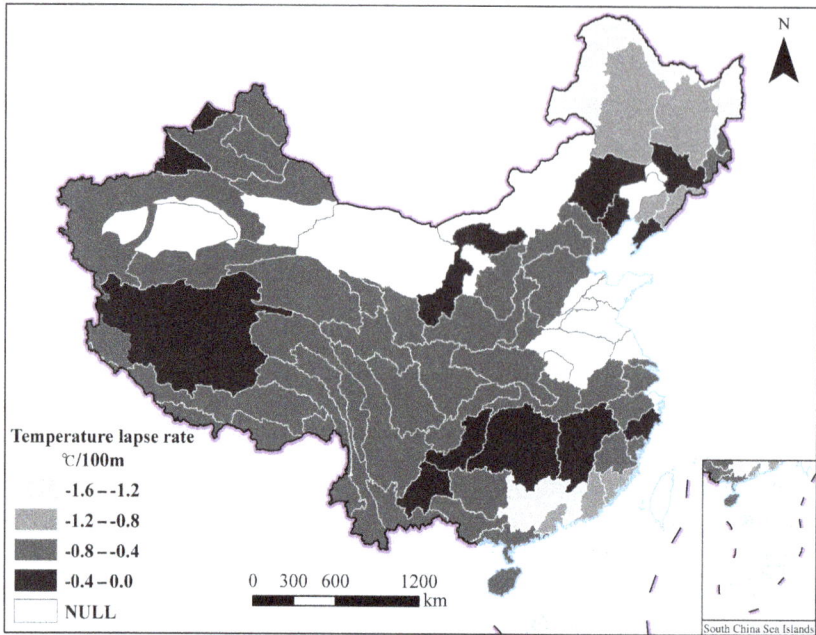

FIGURE 6.3 Change rates of annual precipitation with elevation in Class II WRRs. NULL indicates that the linear correlation between mean annual precipitation and air temperature, and elevation was non-significant

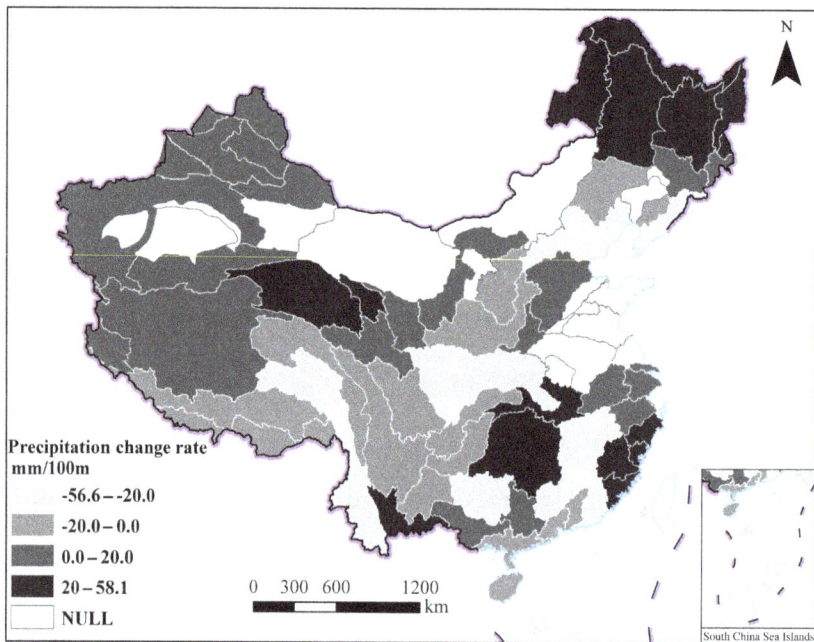

FIGURE 6.4 Change rates of annual air temperature with elevation in Class II WRRs. NULL indicates that the linear correlation between mean annual precipitation and air temperature, and elevation was non-significant.

6.2 A MULTISCALE SPATIAL PARAMETERIZATION SCHEME FOR DESCRIBING DIFFERENCES IN RUNOFF GENERATION

6.2.1 Spatial Parameterization Scheme Design

The CWAM parameters mainly included vegetation parameters, soil parameters, hydrogeological parameters, confluence parameters, and snowmelt parameters. For large-scale modeling, a suitable spatial parameterization scheme should satisfy two requirements: (i) to be able to portray the spatial variability of the runoff generation process in a large-scale region; and (ii) to facilitate the initialization and optimization of the parameters in order to improve the work efficiency. Therefore, a multiscale spatial parameterization scheme of "climatic zones nested water resources regions - sub-basins - contour belts - mosaic plots" was proposed in this study, as shown in Figure 6.5.

6.2.1.1 Climatic Zones Nested Water Resources Regions

On the large-scale regions, the country spans as many as 10 climatic zones from south to north, including the tropical, subtropical, warm temperate, and cold temperate zones. The water and energy conditions in these climatic zones vary considerably, affecting the distribution pattern of vegetation and ecosystems. Besides, the same climatic zone involves several independent river basins, where the size, shape, and characteristics of the water system determine the role of the basins in terms of storage. China consists of 10 Class I WRRs,

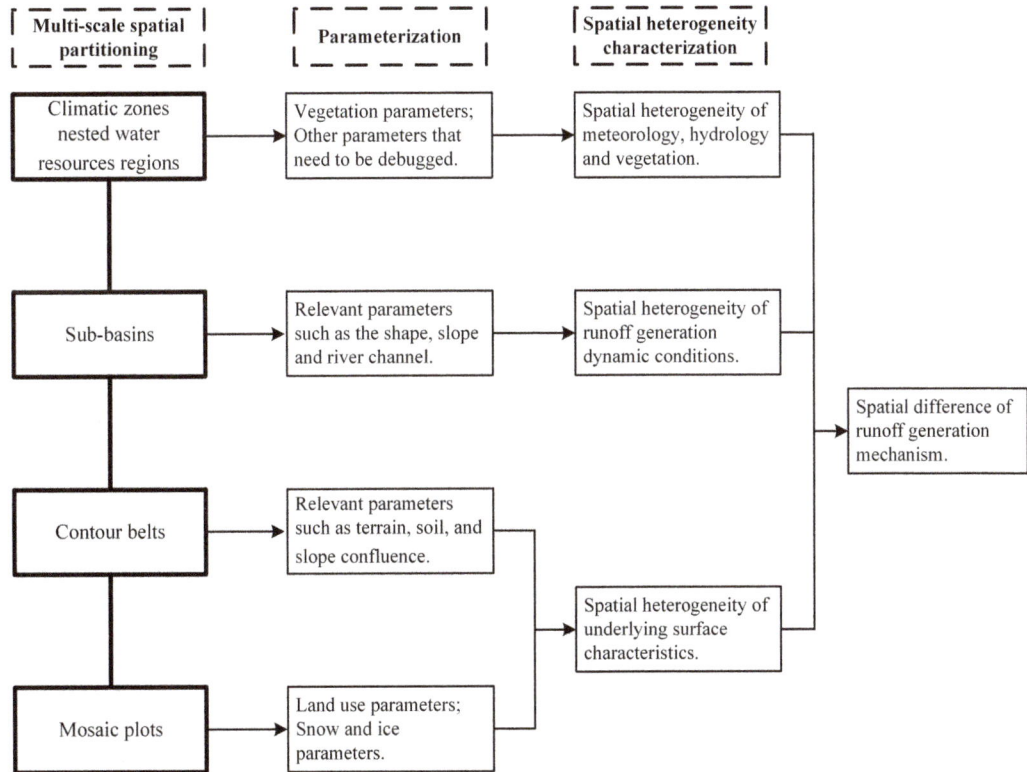

FIGURE 6.5 Schematic diagram of multiscale spatial parameterization scheme.

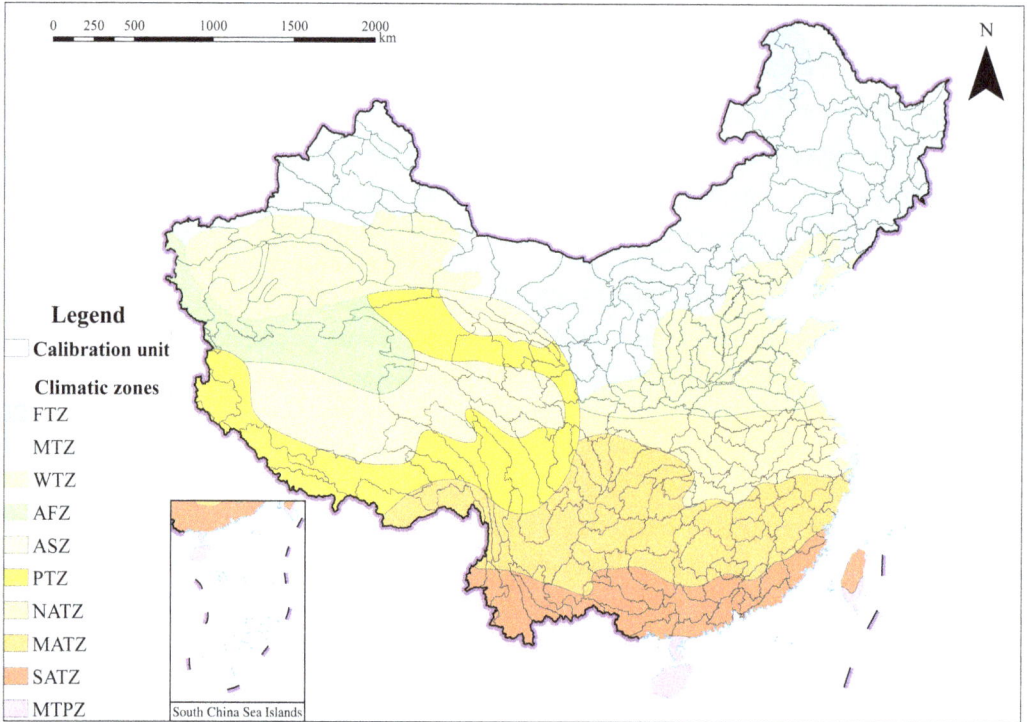

FIGURE 6.6 Distribution of 349 climatic zones nested water resources regions.

80 Class II WRRs, and 210 Class III WRRs. Climatic zones mainly reflect spatial differences in hydrological conditions, whereas water resource regions represent differences in the characteristics of water resources. As a result, considering spatial heterogeneity of meteorological hydrology and underlying surface conditions, China's 10 climatic zones and 210 Class III WRRs were nested to form 349 zones, as shown in Figure 6.6. These zones were used as the parameter tuning units in the CWAM. Meanwhile, the vegetation parameters were assigned in the 349 zones.

6.2.1.2 Sub-basins

Model parameters at the sub-basin scale mainly included the area, slope, elevation, latitude, and longitude, and upstream and downstream topological relationships of the sub-basins, as well as river length, mean specific drop of the river channel, cross-section area of the river channel, and maximum channel depth. These parameters reflected the spatial variability of the dynamical conditions affecting runoff production in a river basin. To quantify channel surface roughness, the channel Manning coefficient was introduced. Besides, the permeability coefficient and the thickness of the riverbed infiltration medium were used as commissioning parameters, considering the exchange between river water and groundwater.

6.2.1.3 Contour Belts

The contour belts are the basic computation units of the CWAM. The area, length, width, slope, and mean elevation of each contour zone were calculated. The main model parameters at the contour belts scale were soil parameters and slope Manning coefficient, reflecting the variability in subsurface conditions. The soil parameters were assigned in each contour belt, and the single soil type of the unit and its corresponding parameters were determined according to the principle of maximum area.

6.2.1.4 Mosaic Plots

In the CWAM, the geomorphological variability within each contour belt was represented by mosaic plots (i.e., 1 km × 1 km). Each mosaic plot corresponds to a land use type. The model classified the subsurface into five groups: soil-vegetation group, non-irrigated farmland group, irrigated farmland group, water body group, and impervious area group. Therefore, the proportion of the area of these groups within the contour belt is counted based on land use raster data. Besides, different land use groups were assigned a corresponding set of snowmelt parameters.

6.2.2 Determination of the Values of the Main Parameters

6.2.2.1 Vegetation Parameters

The MODIS datasets LAI and NDVI from 2001 to 2017 were used to statistically calculate the national 1 km × 1 km raster multi-year average month-by-month leaf area index and vegetation cover. Since the LAI product has a time resolution of 15 d, the first data of each month was chosen as the value for that month. The values of FVC for the 12 months of the year were obtained by the dimidiate pixel model:

$$FVC_i = \frac{NDVI_i - NDVI_{soil}}{NDVI_{veg} - NDVI_{soil}} \tag{6.4}$$

where FVC_i is the fractional vegetation cover in the ith month; $NDVI_{veg}$ denotes the NDVI value at the time of complete vegetation cover; $NDVI_{soil}$ denotes the NDVI value in the bare soil area, which is close to 0 theoretically, but due to the change of surface moisture conditions by atmospheric influences, the $NDVI_{soil}$ varies in time, generally between −0.1 and 0.2.

The results showed that the annual LAI gradually increased from the northwest inland to the southeast coastal areas. In northern China, annual LAI was generally lower than 1.2, while it was between 1.2 and 2.8 in southern China. Moreover, the whole country was further divided into 23 elevation zones using a 250 m elevation gradient. The average NDVI values, calculated in each elevation zone, showed significant vertical zonal heterogeneity. Overall, the value showed a decreasing trend of about 0.1 per 1000 m with respect to increasing elevation. Moreover, annual FVC had a similar distribution to annual LAI. Correlation between these two parameters was significant with increasing elevation, with a Pearson correlation coefficient of 0.93.

The study focused on the impact of elevation on LAI and FVC and incorporated it into the model. Using the monthly MODIS data from 2001 to 2017, the multi-year mean monthly LAI and FVC were computed on a 1 km × 1 km grid over China. Furthermore, the vegetation data of each contour belt was calculated. During the whole simulation period, monthly vegetation maintained the same spatial distribution.

6.2.2.2 Soil Parameters

In the CWAM, soils are classified into four types: sand, loam, clay loam, and clay. The soil moisture characterization parameters of each soil type were determined, as shown in Table 6.3. For example, the distribution of soil saturated water content and soil saturated hydraulic conductivity in China is shown in Figure 6.7.

TABLE 6.3　Soil Moisture Characterization Parameters

Parameters	Sand	Loam	Clay Loam	Clay
Soil Saturated Water Content (cm³/cm³)	0.4	0.466	0.475	0.479
Field Water-Holding Capacity (cm³/cm³)	0.174	0.278	0.365	0.387
Withering Coefficient (cm³/cm³)	0.077	0.120	0.170	0.250
Soil Residual Moisture Content (cm³/cm³)	0.035	0.062	0.136	0.090
Soil Saturated Hydraulic Conductivity (cm/s)	2.5E-3	7.0E-4	2.0E-4	3.0E-5
Havercamp Parameter α	1.75E10	6451	3.61E6	6.58E6
Havercamp Parameter β	16.95	5.56	7.28	9.00
Mualem Parameter n	3.37	3.97	4.17	4.38
Soil Suction Head at Wetting Front SW (cm)	6.1	8.9	12.5	17.5

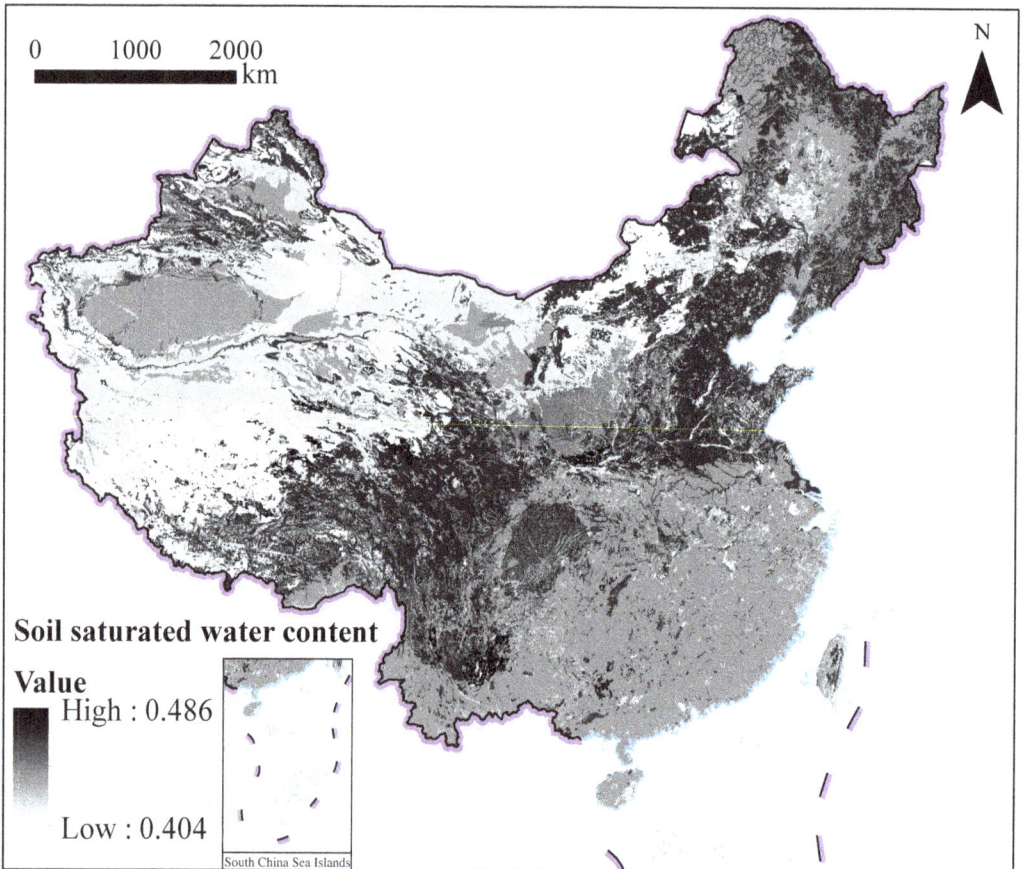

FIGURE 6.7　Distribution of soil saturated water content in China.

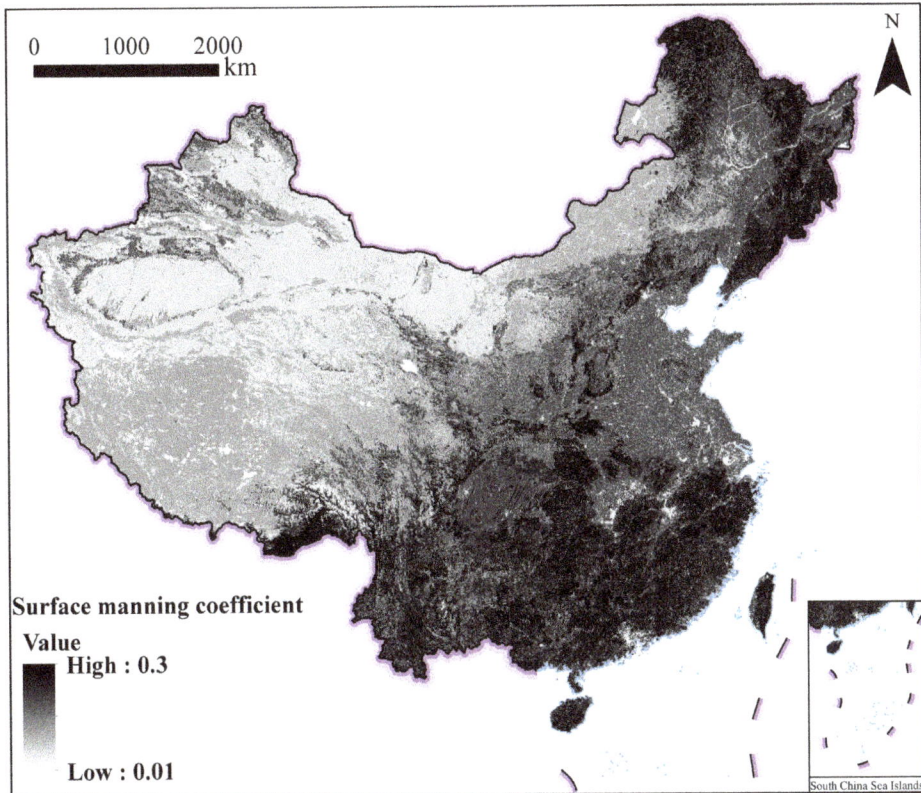

FIGURE 6.8   Distribution of slope Manning coefficient in China in 2015.

6.2.2.3 Runoff Confluence Parameters

In general, confluence includes slope confluence and channel confluence, which mainly involve slope Manning coefficient, channel Manning coefficient, and bed material permeability coefficient. The slope Manning coefficient was taken as an area-weighted average of the Manning roughness for each land use type within the contour belt. Of these, values of 0.3 were taken for forests, 0.1 for grasslands, 0.2 for agricultural land, 0.05 for bare land, 0.02 for habitat sites, and 0.01 for waters/snow. For example, the slope Manning coefficient in 2015 is shown in Figure 6.8. Besides, the initial value of the channel Manning coefficient refers to the previous research results, and the default value is taken as 0.01. The initial value of the quotient of the permeability coefficient of the bed-permeable medium divided by the thickness of the bed-permeable medium was set at 1.0×10^{-5}/s for mountain rivers and 2.5×10^{-5}/s for rivers in the plains.

6.2.2.4 Hydrogeological Parameters

Key hydrogeological parameters required for the CWAM included aquifer thickness, field capacity, and aquifer conductivity. These parameters were evaluated by reviewing papers, reports of the National Hydrogeological Census of County Districts, etc., and include definitive values (borehole data, measured data), empirical values, and assigned values based on lithological distribution. Finally, the parameters were counted on 349 tuning zones, as shown in Figures 6.9–6.11.

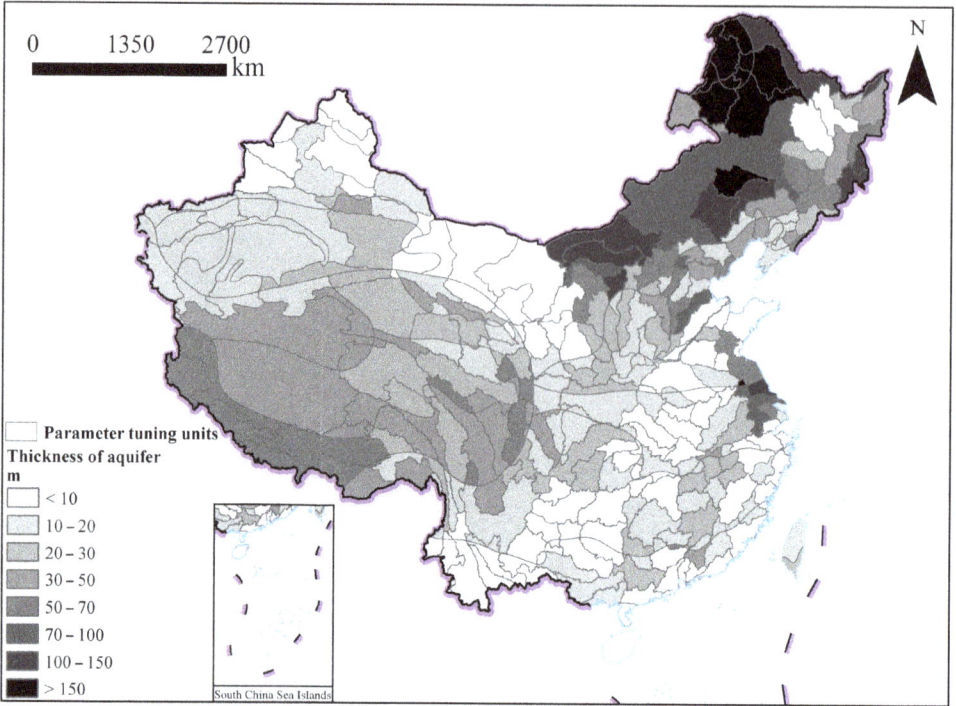

FIGURE 6.9   Distribution of aquifer thickness parameter in China.

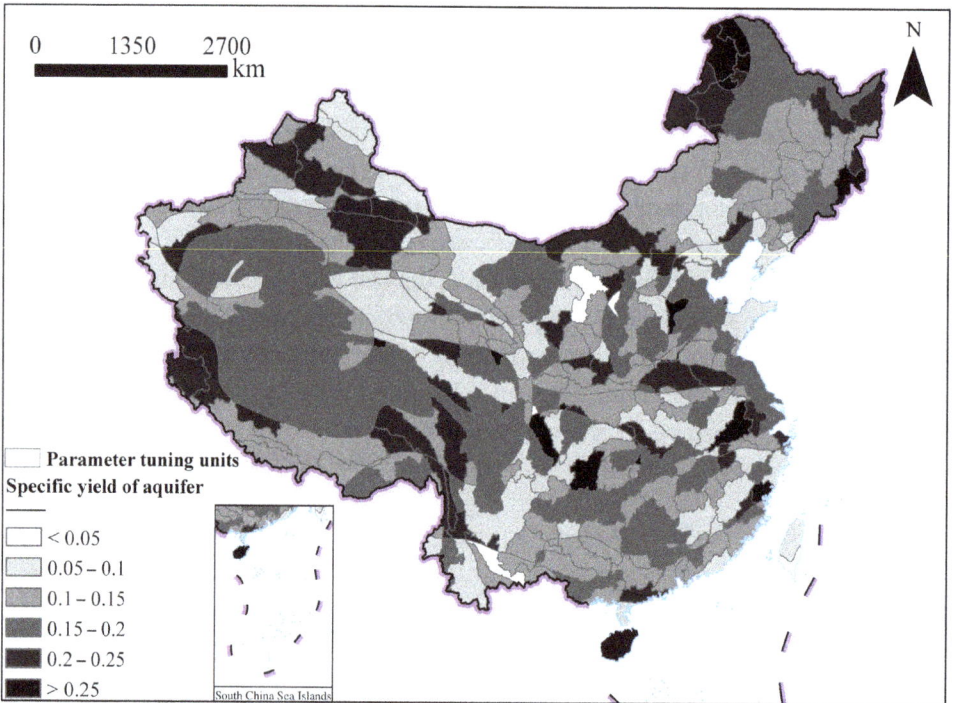

FIGURE 6.10   Distribution of field capacity parameter in China.

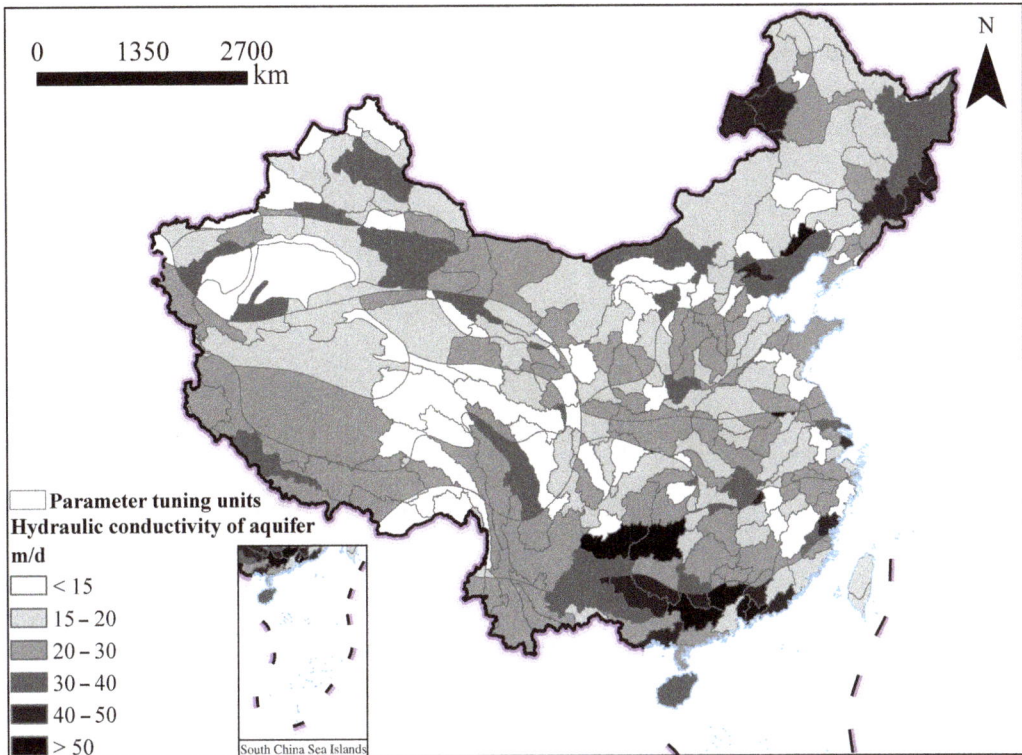

FIGURE 6.11   Distribution of aquifer conductivity parameter in China.

6.3 MODEL CALIBRATION AND VALIDATION

6.3.1 Model Calibration and Validation

Using the CWAM, continuous simulations of 62 years (1956–2017) were conducted in 19,406 sub-basins and 81,687 contour belts nationwide for natural hydrological processes. For model calibration and validation, the simulated monthly streamflow needs to be compared with the statistical ones, which are the traditional naturalized river streamflow by reverting statistical water consumption to the observed ones. To ensure data reliability, the results of the second recent national water resources survey, completed in 2004, were used. The data were the naturalized monthly streamflow of hydrological stations from 1956 to 2000. The following parameters for model calibration were obtained using the sensitivity analysis: maximum depression storage depth of land surface, soil saturated hydraulic conductivity, permeability of riverbed material, Manning roughness, snow melting coefficient, and critical air temperature for snow melting. The calibration was performed on a basis of "try and error".

Consequently, a set of 203 hydrological stations was selected throughout the country. They span over a wide range of hydrological regimes, basin areas, elevations, vadose zone structures, and climatic conditions (see Figure 6.12). Among them, 23, 43, and 17 representative hydrological stations were located in the Loess Plateau, karst region, and cold region, respectively. The calibration period was chosen as 1956–1980, and the validation period

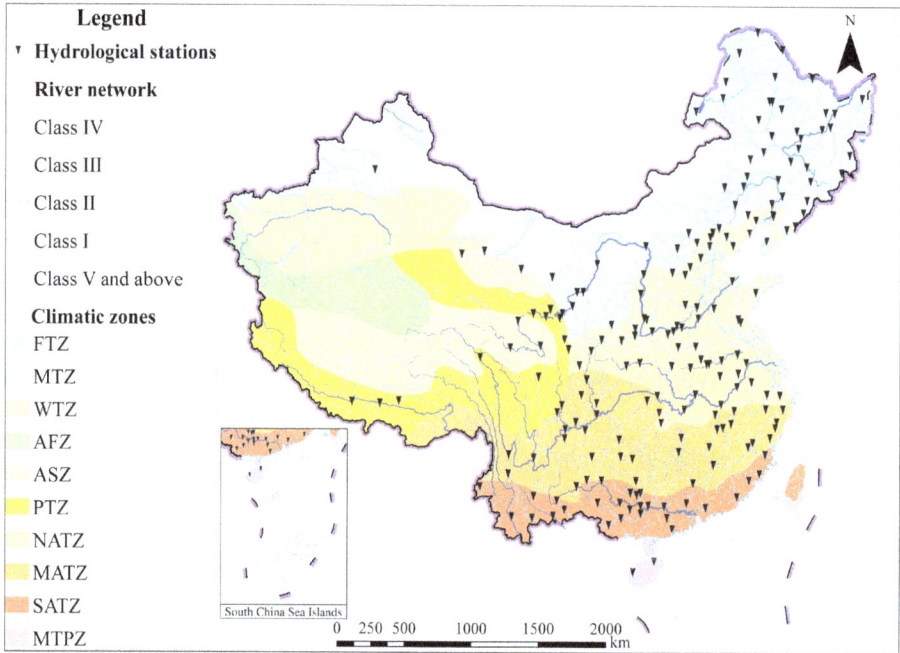

FIGURE 6.12   Spatial distribution of 203 hydrological stations selected.

was 1981–2000. Two widely used model validation criteria were used: minimizing the relative error (*RE*) of annually averaged river runoff, and maximizing the Nash–Sutcliffe efficiency (*NSE*) of monthly streamflow.

6.3.2 Model Applicability in Runoff Processes in China

Calibration and validation results of simulated monthly streamflow at 203 hydrological stations nationwide are reported in Table 6.4. For the calibration period, *RE* ranged from

TABLE 6.4   Calibration and Validation Results of 203 Hydrological Stations in China

WRRs	Total Number of Stations	Calibration Period (1956–1980)				Validation Period (1981–2000)			
		NSE > 0.7		|RE| < 10%		NSE > 0.7		|RE| < 10%	
		NUM	PCT	NUM	PCT	NUM	PCT	NUM	PCT
China	203	165	81%	184	91%	163	80%	193	95%
SRB	28	20	71%	26	94%	18	63%	26	94%
LRB	17	12	73%	17	100%	14	82%	17	100%
HRB	16	14	89%	12	78%	12	75%	14	89%
YRB	31	22	71%	29	93%	22	71%	29	93%
HURB	12	11	90%	10	80%	8	70%	12	100%
YZRB	47	45	96%	41	88%	47	100%	43	92%
SERB	12	12	100%	12	100%	12	100%	12	100%
PRB	25	25	100%	20	80%	23	90%	20	80%
SWRB	10	6	64%	8	82%	7	73%	10	100%
NWRB	5	3	67%	3	67%	3	67%	3	67%

−10.9% to 11.6%, and *NSE* was in the range of 0.66–0.94. There were 184 stations with the absolute *RE* value less than 10%, accounting for 91% of the total number of stations. The number and its proportion in the total number of stations with *NSE* > 0.7 were 165 and 81%, respectively. For the validation period, *RE* ranged from −11.3% to 13.4%, and *NSE* had a range of 0.61–0.96. Among them, the absolute *RE* value of 193 stations was less than 10%, accounting for 95% of the total number of stations. The number and its proportion in the total number of stations with *NSE* > 0.7 were 163 and 80%, respectively.

Spatially, the simulation results were analyzed based on the validation period. Figure 6.13 shows the simulation results of monthly natural streamflow at national hydrological

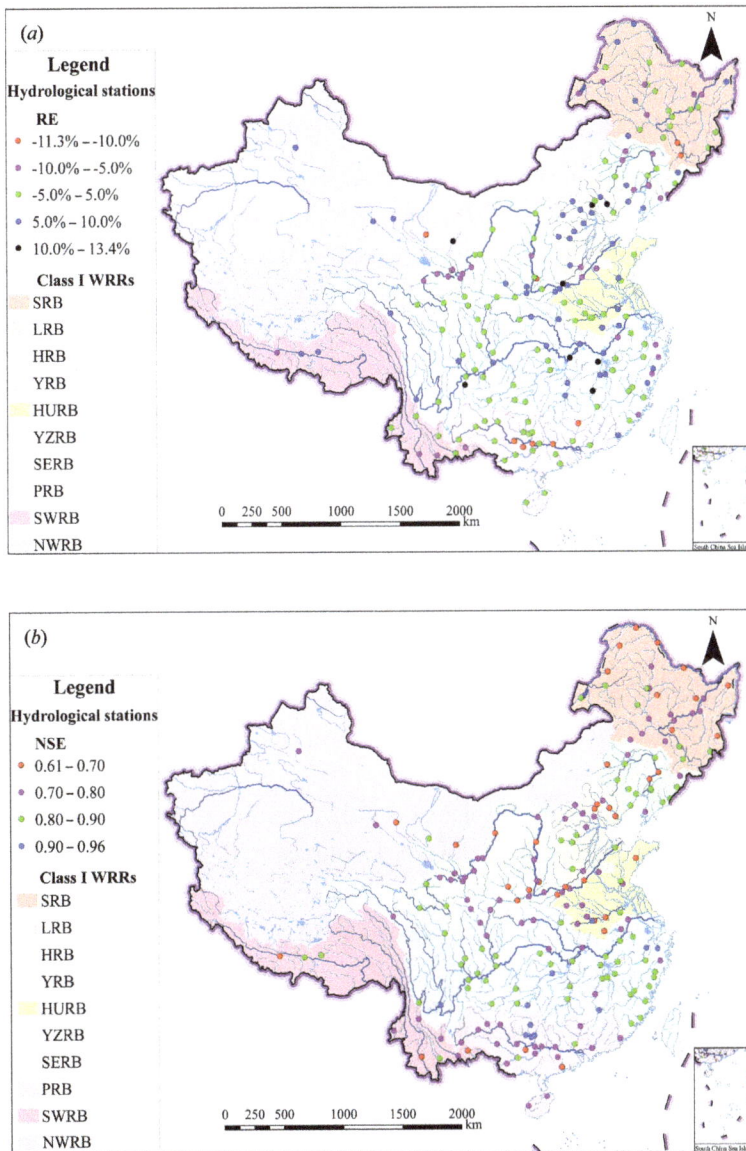

FIGURE 6.13 Schematic illustration of model validation results: (*a*) RE, (*b*) NSE.

stations in the validation period. In southern China, *NSE* was mostly above 0.8, and *RE* was controlled between −5% and 5%, indicating a good simulation effect. *NSE* was mainly between 0.7 and 0.8 in northern China, and the simulated runoff in most stations was higher than its measured value. A possible reason is that the model is insufficient to describe the phenomenon of soil thickening caused by the decrease of groundwater level in the northern region, which needs to be further strengthened. In the Huang-Huai-Hai plain area, the runoff and confluence processes are extremely complex, so the simulation accuracy was relatively lower than in other areas, with *NSE* around 0.7.

In general, the high *RE* and *NSE* values showed that the CWAM can provide efficient simulation in different hydrological regimes, basin areas, elevations, vadose zone structures, and climatic conditions across the country.

6.3.3 Model Applicability in Runoff Processes in China

The results of the second national water resources survey of China (1956–2000 data series) were used to evaluate the performance of the model in assessing water resources, using Class III WRRs as statistical units. Considering that some of the Class III WRRs are located in the desert and have almost no available water resources, they are not involved in the comparative analysis of this study. As a result, a total of 204 Class III WRRs were selected. First, the results of manual statistical surveys of water resources in these Class III WRRs were determined, including surface water resources, groundwater resources, and total water resources. Second, the simulated values of surface water resources, groundwater resources, and total water resources for these regions were obtained from the CWAM. Finally, the correlation coefficient R^2 and the relative error *RE* were chosen to assess the accuracy of the model in simulating the amounts of water resources in China. From the results listed in Table 6.5, the correlation coefficients between the simulated and surveyed values of water resources of 204 Class III WRRs in the country for the period 1956–2000 exceeded 0.7. At the national level as a whole, the relative errors of the annual average surface water resources, underground water resources, and total water resources of the country were −6.85%, 17.5%, and −2.14%, respectively.

The study further statistically analyzed the distribution of relative errors between the simulated and surveyed values of total water resources of 10 Class I WRRs, and the results

TABLE 6.5   Comparative Analysis of the Simulated and Surveyed Values of Water Resources of China

Water Resources	Simulated Values (billion m³)	Surveyed Values (billion m³)	*RE*(%)	R^2
Surface Water Resources	2485.5	2668.3	−6.85	0.82
Groundwater Resources	947.3	806.2	17.5	0.73
Total Water Resources	2710.6	2769.9	−2.14	0.78

TABLE 6.6   Comparative Analysis of the Simulated and Surveyed Values of Water Resources of Class I WRRs

Total Water Resources	SRB	LRB	HRB	YRB	HURB	YZRB	SERB	PRB	SWRB	NWRB
Simulated Values (billion m³)	149	50	37	71	92	996	199	472	578	127
Surveyed Values (billion m³)	145	48	34	74	99	973	193	427	597	120
RE(%)	−2.61	−2.81	−7.30	4.95	7.97	−2.30	−2.67	−9.61	3.32	−5.97

are presented in Table 6.6. It can be seen that the simulation errors of the CWAM on the 10 Class I WRRs were basically controlled between −10% and +10%, and the model had good spatial adaptability.

REFERENCE

Maidment, D.R., 1992. Handbook of Hydrology. McGraw-Hill, New York.

CHAPTER 7

Spatiotemporal Patterns of Hydrological and Water Resources Variables in China

7.1 RESEARCH BACKGROUND

Numerous studies have attempted to quantify and predict the changing characteristics of key hydrological and water resource variables dominated by precipitation, evaporation, and runoff. However, the traditional hydrological research is often focused on the catchment scale. In recent years, increasing pressure on water use at the national and even global scales has forced people to take a broader spatial perspective to advance the understanding of hydrological processes and water resource changes. Two main approaches, the statistics-based and model-based methods, are widely used to understand the large-scale water cycle and its variables. The statistics-based method generally uses satellite remote sensing data on precipitation, evaporation, and water resources at the continental and even global scales. For example, a great number of satellite/reanalysis-based precipitation products were applied in global meteorological and hydrological analyses. The applications of the Gravity Recovery and Climate Experiment (GRACE) satellites revealed regional depletion of groundwater resources (Gleeson et al., 2012). Despite the advantages of these satellite products over gauge-based observations in terms of spatial coverage, they have large uncertainties and errors, arising from sensor deficiencies, retrieval algorithms, and discordant data resolution. The model-based methods mainly rely on the large-scale hydrological modeling, which has made great progress and turned into an interesting field. Numerous large-scale land surface models emerged to investigate changes in hydrological and water resources variables, including the variable infiltration capacity (VIC) (Liang and Xie, 2001), WaterGAP (Alcamo et al., 2003), WASMODM (Widén-Nilsson et al., 2007), the mesoscale Hydrologic Model (mHM) (Kumar et al., 2013), and E-HYPE (Donnelly et al., 2016). However, the grid with a resolution of tens or even hundreds of kilometers was usually used as the computation unit in these models, which distorted the river network,

DOI: 10.1201/9781003646648-7

flow path, and land cover. Consequently, it fails to achieve a refined understanding of the spatiotemporal characteristics of hydrological and water resource variables in large-scale areas. Therefore, many researchers attempted to improve computation units and parameters of the existing physically based hydrological models to extend their applications from the small-/medium- to large-scale regions (Jia et al., 2006; Schuol et al., 2008).

China covers a land area of about 9.6 million km², encompassing many river basins and various climatic zones. However, the hydrological and water resources variables in China have not yet been comprehensively investigated, due to the lack of a detailed national model and operating environment. There were some studies on a single variable of China, such as estimating actual evapotranspiration, analyzing reference evapotranspiration, and discussing the spatial patterns of precipitation. Feng et al. (2018) and Bai et al. (2018), respectively, used GRACE data to present the spatiotemporal patterns of groundwater storage over China. Analysis of runoff characteristics on a large scale with different spatiotemporal resolutions is one of the hot topics among scholars. Using the VIC model, Miao and Wang (2020) produced a flux database of the key water cycle variables of China during 1961–2017, including runoff, evapotranspiration, soil moisture, water storage, etc. However, this study only calculated the runoff flux of the nine largest rivers in China, and their low-resolution data constrained the investigation on the spatial heterogeneity of runoff flux. Sun et al. (2021) examined spatiotemporal shifts of evapotranspiration and runoff across more than half region of China, which is unsatisfactory in terms of spatial coverage.

To comprehensively describe and analyze the spatial heterogeneity of hydrological and water resources in China, we have developed a high-resolution, physically based model named China Water Assessment Model (CWAM). In this chapter, we are committed to applying this model and combining statistical methods to answer several questions: ① What are the temporal trends in key hydrological and water resources variables of China in recent decades? ② Where are the spatial differences in these variables among hundreds of water resource regions? ③ What are the drivers and constraints that affect the changes in these variables? Through the simulation of the CWAM, this study effectively realized the high-resolution display of key water cycle variables across the country. These variables include hydrological variables such as precipitation (P), infiltration (Inf), evaporation (E), and runoff depth (R), as well as surface water resources (SWR), groundwater resources (GWR), and total water resources (TWR). The results are a useful supplement to large-scale water science and provide a new reference for the high-resolution analysis of national-scale water cycle variables through model-based methods.

7.2 SPATIOTEMPORAL PATTERNS OF KEY HYDROLOGICAL VARIABLES IN DIFFERENT WATER RESOURCES REGIONS OF CHINA

7.2.1 Mean Spatial Pattern of Key Hydrological Variables

Figure 7.1 demonstrates the spatial distribution of the long-term means (1956–2017) of the key hydrological variables (i.e., precipitation, runoff, infiltration, and actual evapotranspiration) in China. All variables decreased from the southeast coast to the northwest inland, showing regional differences. This is because those southeast areas are situated closer to

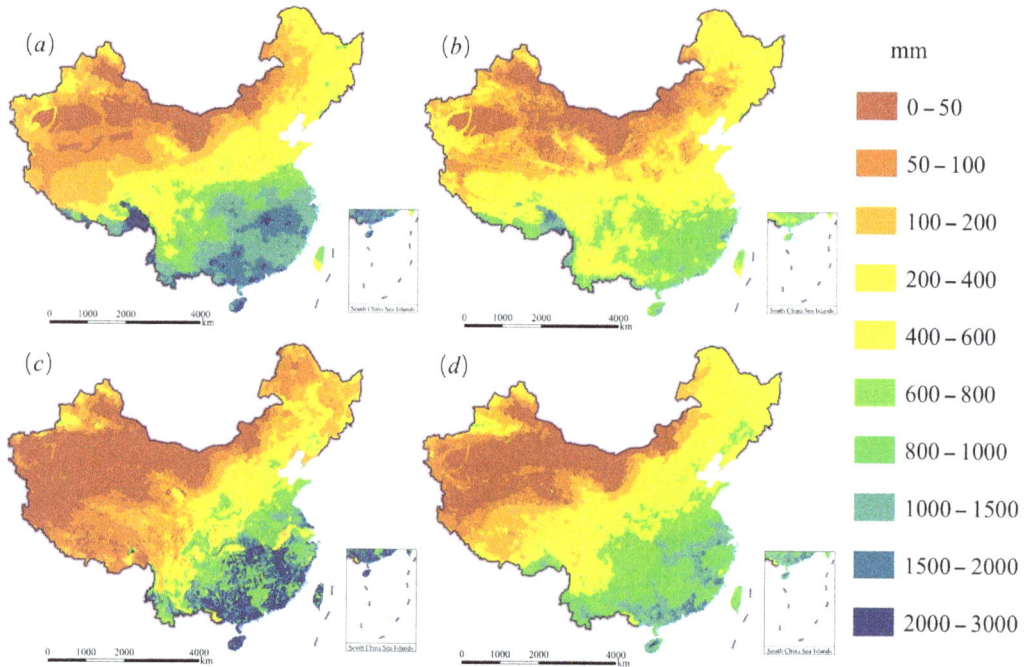

FIGURE 7.1 Mean spatial pattern of hydrological variables in China (1956–2017). (a) P; (b) R; (c) Inf, (d) ET_a.

the coast, compared to the northwest inland, and thus hydrological variables are more likely to be affected by the East Asian monsoon. As seen from Figure 7.1(a), the maximum value area of P was distributed in the southwest regions (SWRB), which is caused by the prevailing Indian monsoon. Compared with the other three variables, Figure 7.1(b) reveals that the area with a small R (<100 mm) in north China is extremely large, which can be explained by that: (i) the regions with R less than 100mm overlap the regions with P less than 800mm, as P is the majority factor restricting R, and (ii) most of the P is mainly infiltrated and evaporated, due to low soil moisture, small air humidity, and mechanism of runoff yield dominated by excess infiltration in the north.

Comparing Figure 7.1(c) with Figure 7.1(d), in the area in northwest China, Inf and ET_a showed similarities in terms of magnitude and spatial pattern. It can be inferred that in northwest China (NWRB), most of the water infiltrated into the soil is evaporated back to the atmosphere. However, in southeast China, which is covered by lush vegetation, ET_a is larger than Inf. This can be explained by that ET_a not only contains water from the soil layer but also adds intercepted water amount by leaves and trunks of vegetation.

7.2.2 Temporal Changes of Key Hydrological Variables in China and Its WRRs

7.2.2.1 Temporal Trends of Hydrological Variables on the National Scale

Figure 7.2 and Table 7.1 demonstrate the 62-year temporal trend of the key hydrological variables, averaged for the entire country. The temporal trends were analyzed with linear regression and the M-K trend test method. The 62-year average values of P, R, Inf, and

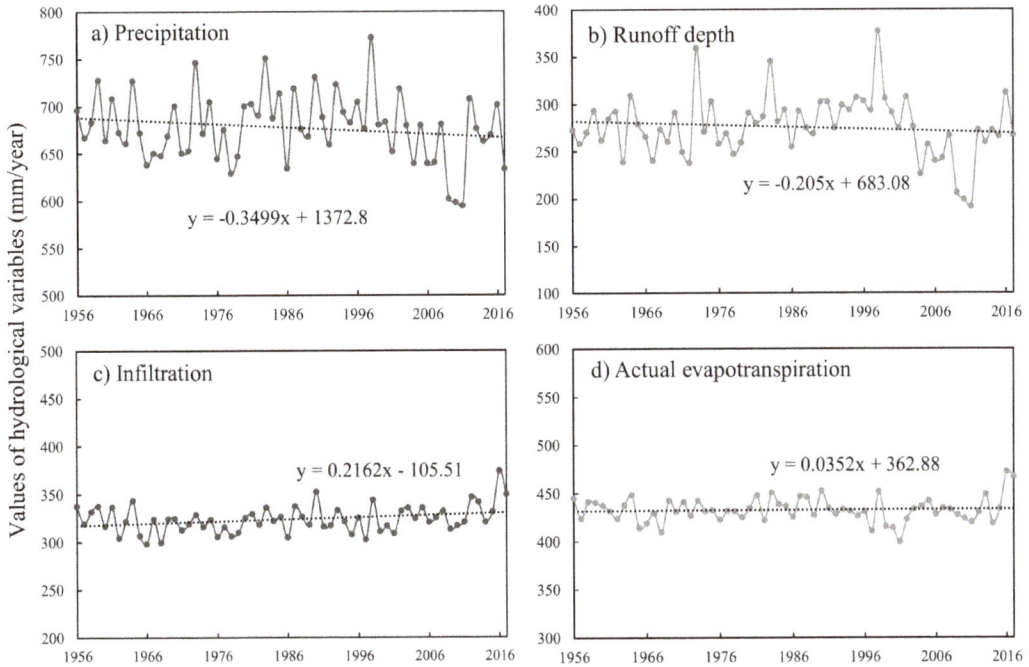

FIGURE 7.2   The linear trend of hydrological variables over China from 1956 to 2017.

ET_a over the country are 678.1 mm, 275.5 mm, 322.6 mm, and 431.6 mm, respectively. Variables P and R have decreasing trends over 62 years, while the Inf and ET_a have weak increasing trends. The M-K test suggested that the trends of the four hydrological variables are not significant. Meanwhile, the four linear regression coefficients were not significantly different from value 0 at the 0.05 level, according to the results of the t-test, indicating that there was no significant trend in the time series of the four variables. The coefficient of variations (CV) of P, R, Inf, and ET_a are 5.24%, 11.80%, 2.57%, and 3.77%, respectively, implying that R is more fluctuating while Inf and ET_a are relatively stable. Meanwhile, R has a significant correlation with P in the time series, which has passed the significance test at the 0.01 level. Annual precipitation decreases at a rate of -0.35 mm/yr (linear trend), suggesting a potentially higher frequency of drought. For example, an extreme drought event in Southwest China that happened in 2009–2010 (years with the lowest precipitation in Figure 7.2(a)) was a "once-a-century drought".

TABLE 7.1   Statistics of Key Hydrological Variables over Time at National Scales

Water Resources	Simulated Values (billion m³)	Surveyed Values (billion m³)	RE(%)	R^2
Surface Water Resources	2485.5	2668.3	−6.85	0.82
Groundwater Resources	947.3	806.2	17.5	0.73
Total Water Resources	2710.6	2769.9	−2.14	0.78

7.2.2.2 Hydrological Variation and Its Differences among WRRs

The trend changes were calculated on the scale of the Class III WRRs, as shown in Figure 7.3. We presented the results on the scale of Class I WRRs. The changes in P across the country are spatially different. The P at most meteorological observation stations in the NWRB shows a significantly increasing trend ($Z_p > 1.96$). The increase in P effectively contributes to the abundance of water resources in NWRB. In addition, regions with increased P also include part of SWRB, downstream of YZRB, some areas in SERB and PRB, but are not significant at the 0.05 confidence level. The regions with a decreasing trend of P ($Z_p < 0$) extend from the northeast to the southwest. These regions specifically include SRB, LRB, HRB, most of YRB, HURB, YZRB midstream, SWRB downstream, and part of PRB. Regions of significant decline ($Z_p < -1.96$) are concentrated in the middle YZRB and scattered in other areas.

From the perspective of changes over the years, there are differences in R of various regions, which can be roughly divided into five areas: ①Most areas of NWRB have a significant increasing trend ($Z_r > 1.96$); ②In the north-central area (parts of YRB and HRB), the changing trend of R show discretized spatial distribution, irregularly switched between increasing and decreasing; ③In the northeast area (SRB and LRB), increasing trends and decreasing trends present clustered distributions, the general pattern is increased in the west and decreased in the east; ④The southeastern area (downstream of YZRB, parts of HURB and SERB) shows an increasing tendency ($0 < Z_r < 1.96$); ⑤Conversely, southwestern

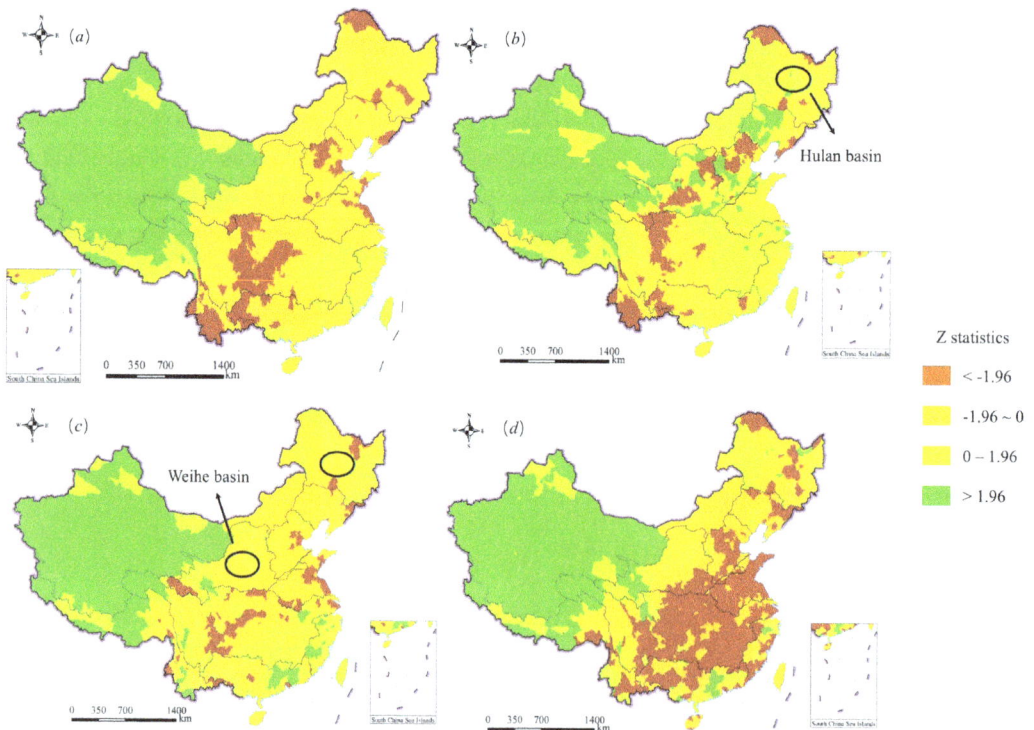

FIGURE 7.3   Spatial pattern of results for the M-K trend test. (a) Z_p; (b) Z_r; (c) Z_{inf}; (d) Zet_a.

areas (up and middle stream of YZRB and PRB; part of SWRB) show clustered decreasing tendency ($Z_r < 0$).

Analyzing the changes in infiltration volume from the inter-annual change trend, the results showed its changing tendency (Figure 7.3(c)) overlaps well with the spatial distribution of Z_p (Figure 7.3(a)). Most of the Z_{inf} in NWRB showed a significant and concentrated increase. The increase and decrease coexisted in other regions of China, which are dominated by insignificant trends. Some scattered basins in the southwestern and eastern regions show significantly decreasing trends.

According to the results of the M-K trend test, the ET_a in China has changed drastically. There are concentrated and large-scale areas with reduced ET_a that are significant (Zet_a < −1.96) from the central to the southeast (i.e., middle of HRB, middle and downstream of YZRB, HURB, upstream of PRB). Most of the northwest (NWRB) shows a significant trend of change that is opposed to that of the central and southeast regions. Other regions show discreetly increasing or decreasing trends, but there were more areas with decreasing trends than increasing within the insignificant range.

Furthermore, the CV values of four key hydrological variables on Class III WRRs were calculated and summarized in the Class I WRRs, as listed in Table 7.2. Overall, the CV values for the four variables in the Class I WRRs showed variability between the north and south. Compared with the south, the CV values for P and ET_a were slightly greater in the north, while the R values were significantly greater. Conversely, the CV values for Inf were greater in the south.

7.2.3 Drivers and Constraints for Changes in Key Hydrological Variables

7.2.3.1 Driving Forces for Changes in R and Inf

Changes in the climate and underlying surface are the main driving forces for variations in runoff and infiltration. The northwestern region presents a spatially continuous and large-scale trend in which both runoff and infiltration increased significantly. This is likely caused by increased P and thawing with warmer temperatures (T).

TABLE 7.2 The CV Values of Four Key Hydrological Variables on Class III WRRs

Class I WRRs	Location	Number of Class III WRRs	P Mean CV	P Median CV	R Mean CV	R Median CV	ET Mean CV	ET Median CV	Inf Mean CV	Inf Median CV
SRB	North China	18	5%	5%	6%	5%	3%	3%	6%	6%
LRB		12	4%	4%	7%	3%	2%	2%	4%	2%
HRB		15	8%	7%	16%	14%	6%	5%	3%	2%
YRB		29	5%	4%	16%	14%	4%	3%	3%	2%
HURB		14	6%	6%	10%	9%	3%	4%	6%	6%
NWRB		33	11%	11%	17%	18%	7%	8%	4%	3%
YZRB	South China	45	3%	3%	4%	3%	2%	1%	8%	6%
PRB		10	3%	2%	2%	2%	1%	1%	8%	6%
SERB		20	3%	3%	2%	3%	2%	2%	8%	7%
SWRB		14	6%	4%	5%	2%	2%	2%	14%	10%

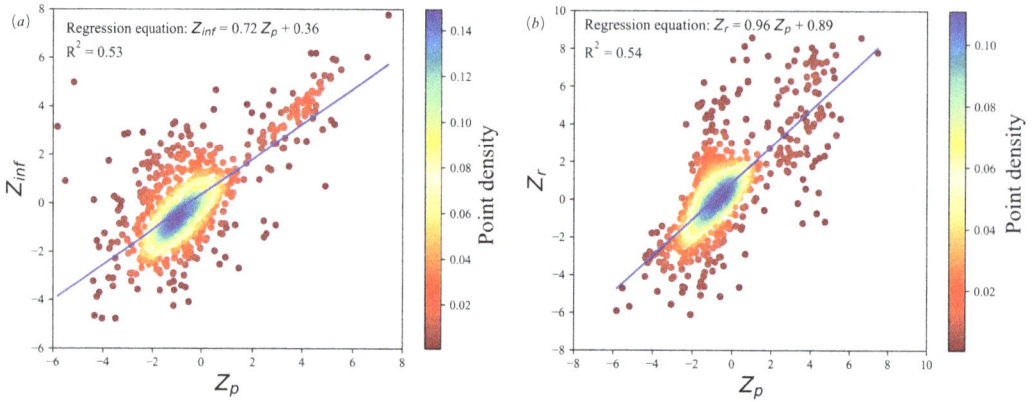

FIGURE 7.4 Linear regression relationship and point density distribution map. *a*) Linear regression relationship between Z_{inf} and Z_p; *b*) Linear regression relationship between Z_r and Z_p.

The size of Z values can reflect the significant degree of change in the variables over time. Therefore, the correlation analysis of the two statistics (Z values) can be used to determine whether there is a correlation between the intensity of change of the two variables. By analyzing the correlation between Z_p and Z_r, we found that they are significantly correlated at the 0.01 level with a linear correlation coefficient of 0.74, i.e., there is a strong positive correlation in the intensity of change between R and P (Figure 7.4(a)). A similar correlation analysis between Z_p and Z_{inf} is also performed here. The result showed that they are significantly correlated at the 0.01 level with a linear correlation coefficient of 0.73, i.e., there is a strong positive correlation in the intensity of change between Inf and P (Figure 7.4(b)).

Climate warming has caused the melting of glaciers and the thawing of frozen soils, which has increased water production in the regions. Although a weak negative correlation between Z_t and Z_r was confirmed when performing spatial correlation analysis using gridded meteorological and hydrological data, meaning that there is no direct evidence that warming is causing an increase in water production. However, a positive effect of temperature on R and Inf still cannot be ruled out. Many studies (Lei et al., 2017; Luo et al., 2018; Gao et al., 2021) reported that water yield in this region has increased significantly with the melting of glacier snow in recent years. Conversely, it is unlikely that this is a change caused by human activity since the vegetation coverage and population density in northwestern China are much lower than in other regions. There are also the southeast coastal areas (downstream of YZRB, parts of PRB, and SERB) affected by the P change ($0 < Z_p$), where both runoff and infiltration show an increasing tendency ($0 < Z_r, Z_{inf}$).

Another phenomenon is that the driving force of underlying surface change is greater than that of the P variable. In those regions, runoff and infiltration roughly show a reverse trend. This change in the underlying surface is reflected in the forest area, the farming area, and urbanization. A representative river basin is the Hulan Basin (shown in Figure 7.3) in the Northeast of China. With the implementation of the "large granary in the Northeast" strategy in recent years, the agricultural area in the Hulan Basin has expanded rapidly. In 2015, the area of arable land increased by 56.58% compared to 1980. The proportion of forests decreased from 31.89% in 1980 to 28.34% in 2015, which directly increased the runoff and reduced the infiltration capacity. Another basin is the Weihe Basin (shown in Figure 7.3), located in the middle of the YRB, where soil and water conservation have been implemented for many years. The area of vegetation, such as forests and grasslands, has increased, accompanied by a decrease in bare land. This change has increased the water conservation capacity of the basin, resulting in a decrease in runoff and an increase in infiltration.

There is a clear difference between climate change and underlying surface variability in scope and intensity. On the national scale, the changes in runoff and infiltration driven by climate change (including precipitation and air temperature) are more noticeable, as evidenced by the significant changes in the northwestern region. In contrast, the underlying surface changes under human activities are limited; only the impact of forest planting (Weihe Basin) and agricultural planting (Hulan Basin) in a scaled area can be identified.

7.2.3.2 Constraints on Actual ET

Figure 7.5 shows the spatial pattern of the long-term mean of potential evapotranspiration (PET), which is the potential rate when sufficient water is available (Kirkham, 2014). The spatial pattern of PET in China shows a spatial pattern opposite to ET_a, meaning that water availability plays a critical role in constraining ET_a. However, this constraint differs

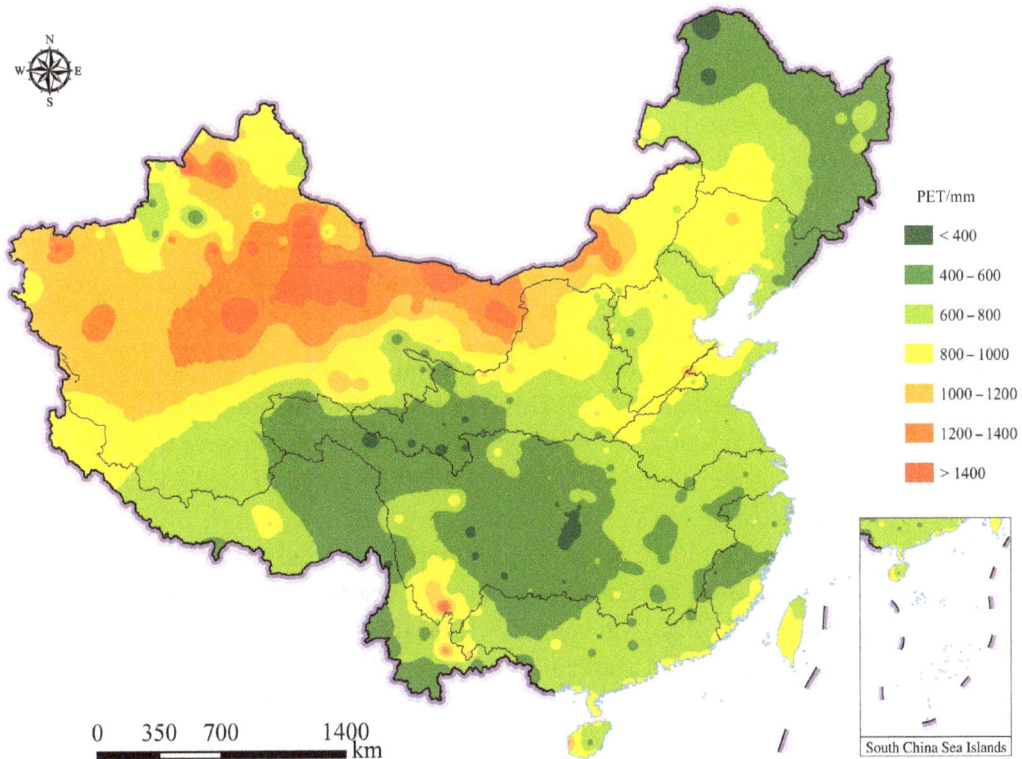

FIGURE 7.5   Spatial distribution of the mean *PET.*

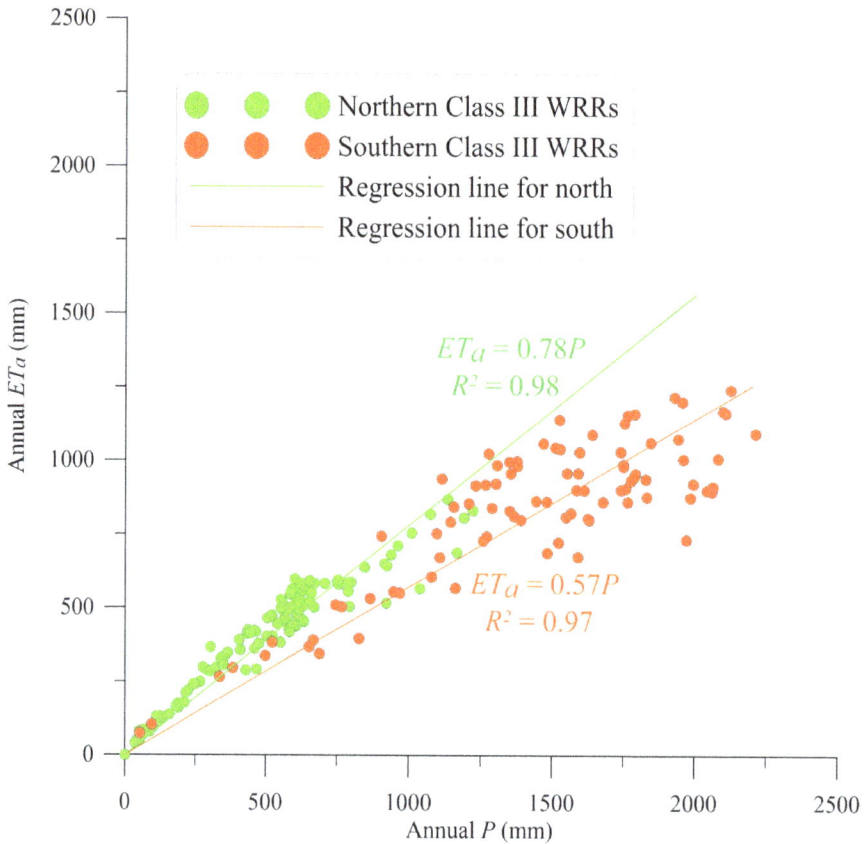

FIGURE 7.6 Linear regression relationship between ET_a and P in north and south Class III WRRs; R^2 is the goodness of fit.

in the southern and northern regions. The results of the T-test for correlation between ET_a and P showed that both variables are significantly correlated at the 0.01 level in the northern (SRB, LRB, HRB, HURB, YRB, and NWRB) and southern (SWRB, YZRB, SERB, and PRB), respectively, over China. However, the regression coefficient (Figure 7.6) is greater in the north (0.78) than in the south (0.57). This means that P in the north is a stronger constraint on ET_a than in the south, which is consistent with the perspective (McVicar et al., 2012) that the north is water-limited ET_a and the south is energy-limited ET_a based on the concept of the Budyko framework.

The significant reduction in large areas of the center region extending to southeastern regions was presumed to be related to a decrease in vegetation cover. Regarding forest evapotranspiration (Figure 7.7(a)), grassland evapotranspiration (Figure 7.7(b)), and surface interception and waterbody evaporation (Figure 7.7(c)), in the above-mentioned ET_a decreasing region, both ET_g and E_w show a significant decreasing trend ($Z<-1.96$), and the range of declines in ET_f is also large. These suggest a possibility that the vegetation cover of the region, particularly the area covered by grassland, has declined significantly over the last few decades. The reduction in E_w (Figure 7.7(c)) should be caused by a reduction in

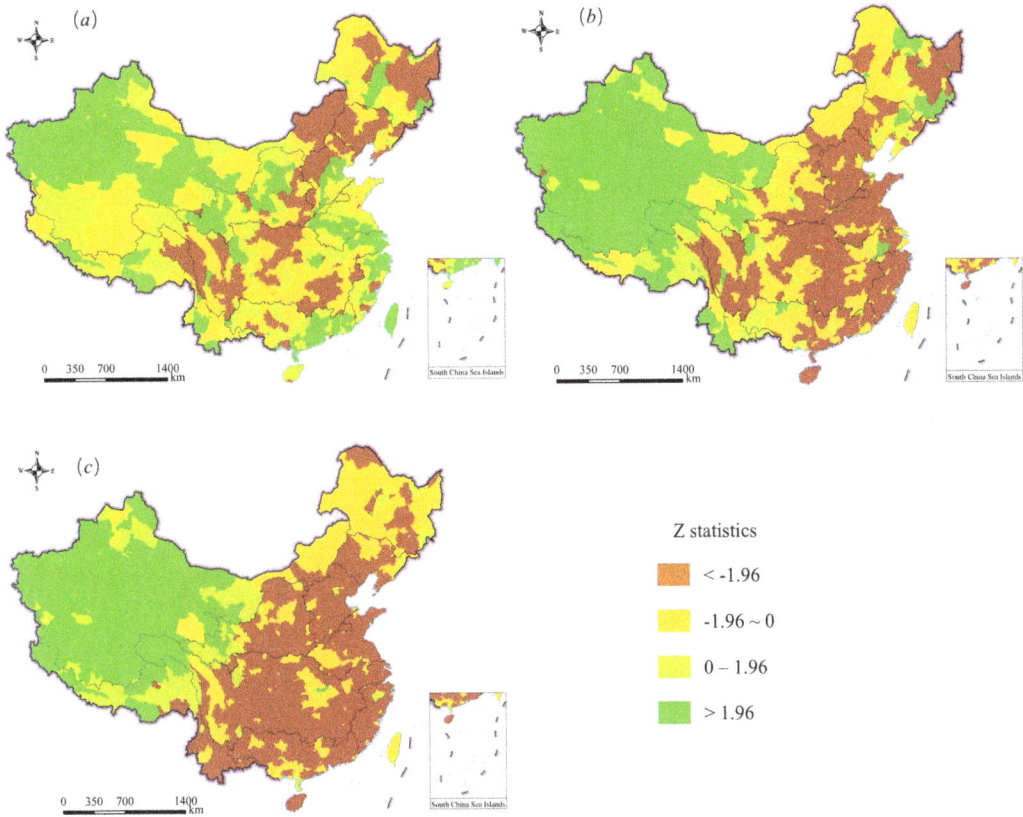

FIGURE 7.7 Spatial distribution of M-K trend test for *a*) ET_f, *b*) ET_g, and *c*) E_w; *ET* is evapotranspiration and *E* is the evaporation; subscripts *f*, *g*, and *w* mean forest, grassland, and surface waterbody, respectively.

surface retention capacity due to changes in vegetation cover. This is consistent with the finding that the vegetation coverage area, especially the grassland in southeastern China, has decreased significantly from 1980 to 2017.

The NWRB has the largest potential evapotranspiration in China. When the waterbody surface evaporation and interception evaporation (Figure 7.7(c)) in this area increase significantly, as well as the forest transpiration also increases to a certain extent, the ET_a of NWRB has a significant increasing trend, which has become an inevitable result. On the one hand, the trend of precipitation increases the amount of water available in NWRB. On the other hand, the melting of many glaciers and snow in the region is another important factor for water availability that cannot be ignored.

7.3 SPATIOTEMPORAL PATTERNS OF WATER RESOURCES IN DIFFERENT WATER RESOURCES REGIONS OF CHINA

7.3.1 Spatial Distribution and Its Changes in Water Resources

We evaluated the amount of water resources during the period 1956–2017 in China based on the output variables of the CWAM. The results show that the long-term mean value of

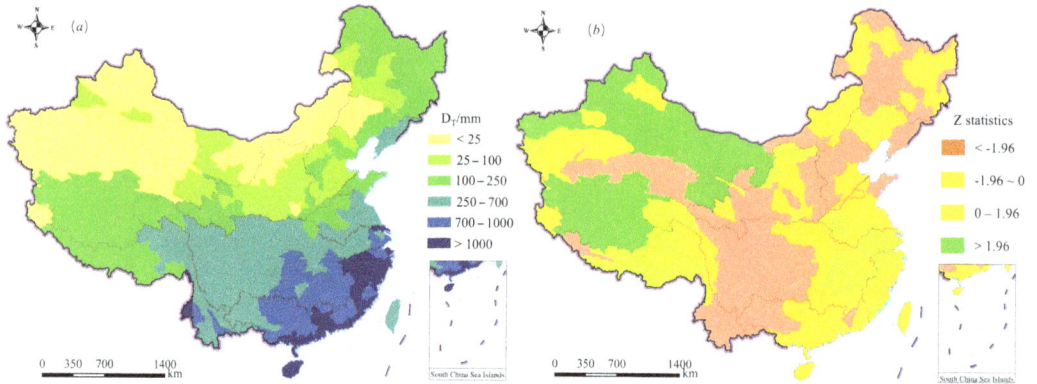

FIGURE 7.8   Spatial distribution of *a*) D_T and *b*) Z values.

TWR in China was $2.81 \times 10^{11} m^3$/year, of which the *SWR* value was $2.63 \times 10^{11} m^3$/year and the *GWR* value was $0.82 \times 10^{11} m^3$/year. It is necessary to note that the average overlap water resources between *SWR* and *GWR* were $0.64 \times 10^{11} m^3$/year, which is the part of the common renewable water resources to both surface water and groundwater (defined in (FAO, 2003), which is groundwater drainage into rivers, typically the base flow of rivers, minus seepage from rivers into aquifers).

In the spatial distribution of the Class III WRRs scale, the depth of *TWR* (D_T) gradually increased from northwest to southeast (Figure 7.8). The values of D_T in the northwestern region and the northern region adjacent to Mongolia were mostly below 100mm. The values of D_T in the northeast region were 100–250 mm, but the values in the coastal areas within it are significantly higher than those of others, up to 700 mm. The southeast region has the most abundant water resources, with an average annual D_T value of about 1000 mm. In terms of time evolution, the values of *TWR* in HRB, SWRB, upper-middle stream of YZRB, and SRB decreased significantly ($Z < -1.96$), likely caused by the combined effects of climate factor and underlying surface change. Conversely, the values of *TWR* in NWRB showed a significant increasing trend ($Z > 1.96$), possibly due to the melting of glaciers and the thawing of frozen soil.

7.3.2 Inter-annual Variability in Water Resources

Table 7.3 shows the Inter-annual variability of water resources during 1956–2017 in China and its Class I WRRs. Based on the whole country and the 10 Class I WRRs, the inter-annual variability of *TWR* from 1956 to 2017 was analyzed, and the results are shown in Figure 7.9. The results showed that, relative to the period 1956–1979, there was a slight increase of about 2% in the national average annual value of *TWR* in the period 1980–2000. Among them, SRB had the largest increase (15%), and HRB had the most significant decrease (40%). The national average annual value of *TWR* showed a decrease (3.5%) in the period 2001–2017. Except for SRB, HURB, and NWRB, all the other Class I WRRs showed varying degrees of decline, with the rate of decline varying between 1.2% and 49.6%.

TABLE 7.3 Inter-annual Variability of Water Resources in China and Its Class I WRRs

Regions	Variability of TWR/%		Variability of SWR/%		Variability of GWR/%	
	Case (1)	Case (2)	Case (1)	Case (2)	Case (1)	Case (2)
SRB	14.51	1.70	12.37	1.88	14.49	0.19
LRB	−10.96	−4.50	−10.22	−3.38	−4.31	−5.94
HRB	−37.40	−49.59	−38.07	−42.04	−31.55	−51.70
YRB	−9.91	−12.32	−9.44	−12.76	2.88	−2.09
HURB	−11.44	9.13	−10.39	8.92	−13.99	9.95
NWRB	7.55	−1.95	6.21	−1.79	8.72	−4.85
YZRB	7.81	−1.51	8.27	−0.83	6.06	−2.48
PRB	1.94	−10.80	1.52	−10.10	−2.33	−8.95
SERB	−1.94	−1.20	−5.43	−1.13	5.80	−1.92
SWRB	−1.25	10.22	2.47	14.22	−6.87	−9.89
China	1.79	−3.45	1.60	−2.59	3.32	−5.21

Note: Inter-annual variability in the period 1980–2000 relative to the period 1956–1979 was represented as Case (1), and the inter-annual variability in the period 2001–2017 relative to the period 1956–1979 was represented as Case (2).

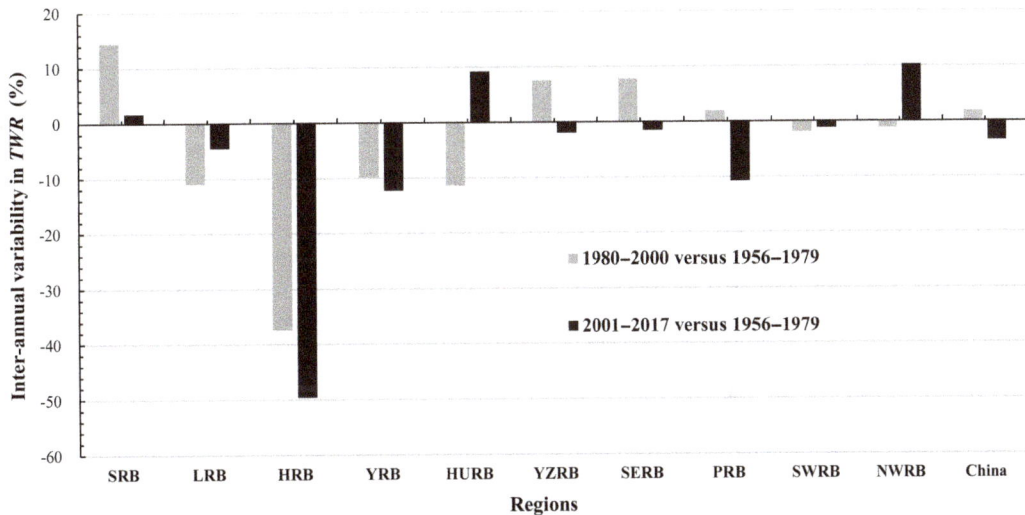

FIGURE 7.9 Inter-annual variability in *TWR* in China and its Class I WRRs.

Relative to the period 1956–1979, the national average annual value of *SWR* increased slightly by 1.6% from 1980 to 2000 (Figure 7.10). Among them, SRB showed the largest increase (12.3%), and HURB showed the most significant decrease (38%). The national average annual value of *SWR* decreased by 2.6% from 2001 to 2017. Of these, NWRB had the largest increase (14%), and HRB had the most significant decrease (40%).

The average annual value of *GWR* in the country increased in 1980–2000 relative to 1956–1979, with an increase of about 3% (Figure 7.11). SRB, YZRB, and SERB showed an

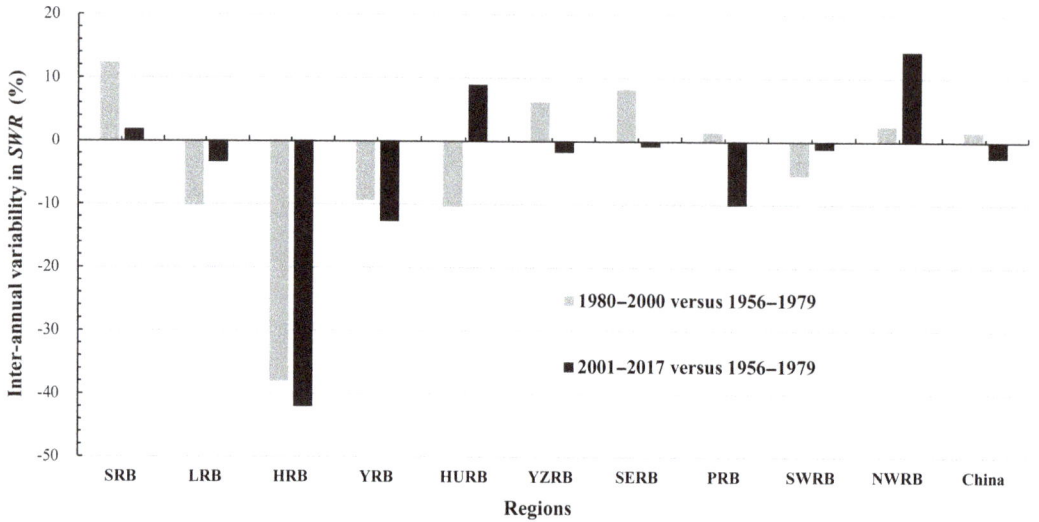

FIGURE 7.10 Inter-annual variability in *SWR* in China and its Class I WRRs.

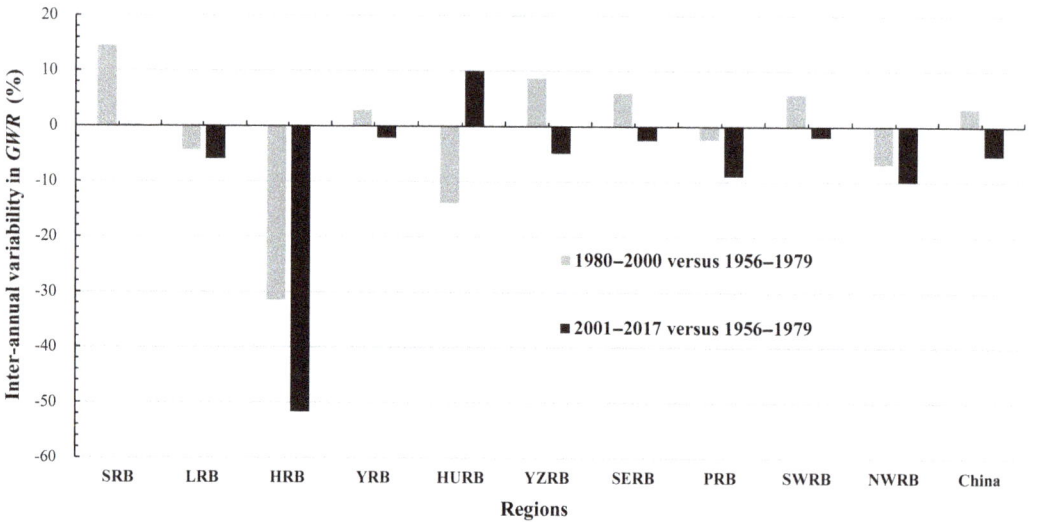

FIGURE 7.11 Inter-annual variability in *GWR* in China and its Class I WRRs.

increasing trend (2%–15%), while HRB showed the most significant decreasing trend, with a 32% decrease. The national average annual value of *GWR* decreased (5%) during 2001–2017. Except for SRB and HURB, the other Class I WRRs showed decreases of varying degrees (2%–52%), with increases of 0.2% and 9.9% in the SRB and HURB, respectively.

7.3.3 Increased Stress on Water Resources Utilization in North China

Figure 7.12 illustrates the mean values of the depth of *TWR* and water consumption on the national scale and Class I WRRs scale for the years 1956–2017. Because of the limitation

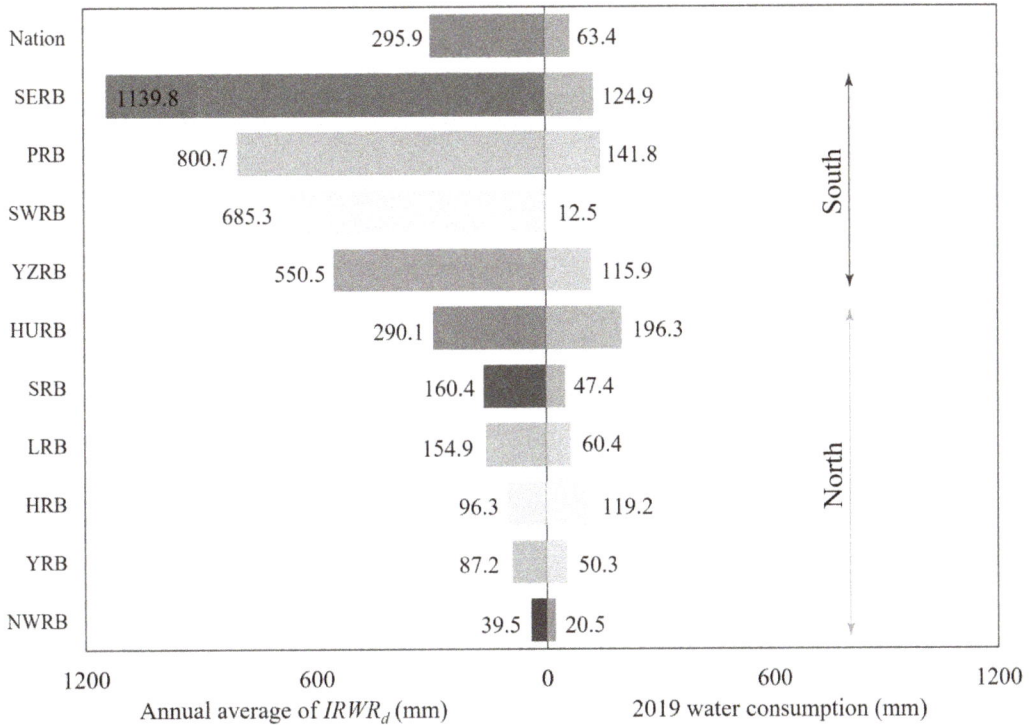

	Annual average of $IRWR_d$ (mm)	2019 water consumption (mm)
Nation	295.9	63.4
SERB	1139.8	124.9
PRB	800.7	141.8
SWRB	685.3	12.5
YZRB	550.5	115.9
HURB	290.1	196.3
SRB	160.4	47.4
LRB	154.9	60.4
HRB	96.3	119.2
YRB	87.2	50.3
NWRB	39.5	20.5

FIGURE 7.12 Annual averages (1956–2017) of TWR and 2019 water consumption.

of available data, the current water demand by Class I WRRs is represented by water consumption in 2019. Despite variances in the development and water use efficiency across WRRs, the data still demonstrate the mismatches between water resources and population distribution in China.

It can be seen that, for North China, the water resources utilization of six Class I WRRs showed different water stress profiles. The most notable feature was a dramatic fall in water resources over more than half a century in HRB, which was driven by a decrease in precipitation. The rate of decline of water resources in the YRB has slowed. However, combined with the water consumption results, this region was still facing greater water stress. Hopefully, the execution of the "High-quality Development of the Yellow River Basin" will break this dilemma. The amount of water resources of HURB showed a pattern of decreasing and then increasing. This unstable pattern of change is highly detrimental to the sustainable utilization of water resources and is prone to extreme drought and extreme flooding events to occur alternately. Although NWRB showed an increase in water resources, to which thawing processes made a non-negligible contribution, NWRB may encounter water shortages as future thawing water decreases. Residents in the NWRB should be extra cautious when engaging in production activities. The reduction of runoff in recent years has contributed significantly to the severe shrinkage of the wetland area in SRB, and the shrinkage of the wetland has fed back into the reduced capacity to store

TABLE 7.4   Changes of Water Resources in China and Its WRRs in 2021–2050 Relative to 1956–2017

| | | | Change in Water Resources Relative to the Base Period (1956–2017) | | | | | |
| | Average Value of Change/% | Number of Class III WRRs | ≤−10% | | (−10%, 10%) | | ≥10% | |
Regions			Number	Proportion /%	Number	Proportion /%	Number	Proportion /%
SRB	5.05	18	0	0.0	15	83.0	3	17.0
LRB	13.67	12	0	0.0	6	50.0	6	50.0
HRB	8.01	15	0	0.0	8	53.0	7	47.0
YRB	1.36	29	0	0.0	25	86.0	4	14.0
HURB	−0.03	14	0	0.0	13	93.0	1	7.0
NWRB	−1.29	45	0	0.0	43	96.0	2	4.0
YZRB	0.79	9	1	11.0	8	90.0	0	0.0
PRB	−1.63	19	1	5.0	16	84.0	3	16.0
SERB	−3.59	14	0	0.0	13	93.0	1	7.0
SWRB	10.09	29	4	14.0	17	59.0	12	41.0
China	2.82	204	6	3.0	165	81.0	39	19.0

runoff, which may result in potential disasters such as flooding. Thus, the runoff storage capacity in SRB requires attention.

7.3.4 Future Changes in Water Resources

The CWAM was used to briefly explore the possible future trends of water resources in China. Meteorological factor changes in the RegCM4 Regional Climate Model (RCM4) Medium Future Emission Scenario (RCP4.5) for greenhouse gases and aerosols were used as inputs to drive the national model to project the changes of water resources in China for the period 2021–2050. The results are shown in Table 7.4. At the national level, there is a slight increase in future water resources relative to the baseline period (1956–2017) under the RCP4.5 emissions scenario, with a relative rate of change of about 2.8%. Among 10 Class I WRRs, HURB and SERB showed small changes in water resources, with average values of −0.03% and 0.79%, respectively. In addition, the water resources of YZRB, PRB, and SWRB had a decreasing trend. In contrast, the northern river basins basically showed an increasing trend, with LRB and NWRB showing the largest increase, over 10% overall, followed by HRB, with an average increase of 8%, and YRB, with a slight increasing trend. Furthermore, for most of Class III WRRs, the variation in water resources relative to the base period (1956–2017) was between ±10%.

Figure 7.13 illustrates the differences in the changes of water resources in different regions due to spatial differences in climate change, using Class III WRRs as statistical units. It can be seen that the spatial distribution of water resource variability in the different WRRs in NWRB and HRB varied considerably, with a difference of 56% and 41% between the maximum and minimum relative rates of change, respectively. In comparison, the spatial differences in changes in Class III WRRs in SERB were the smallest, with relative rates of change ranging from ±5%.

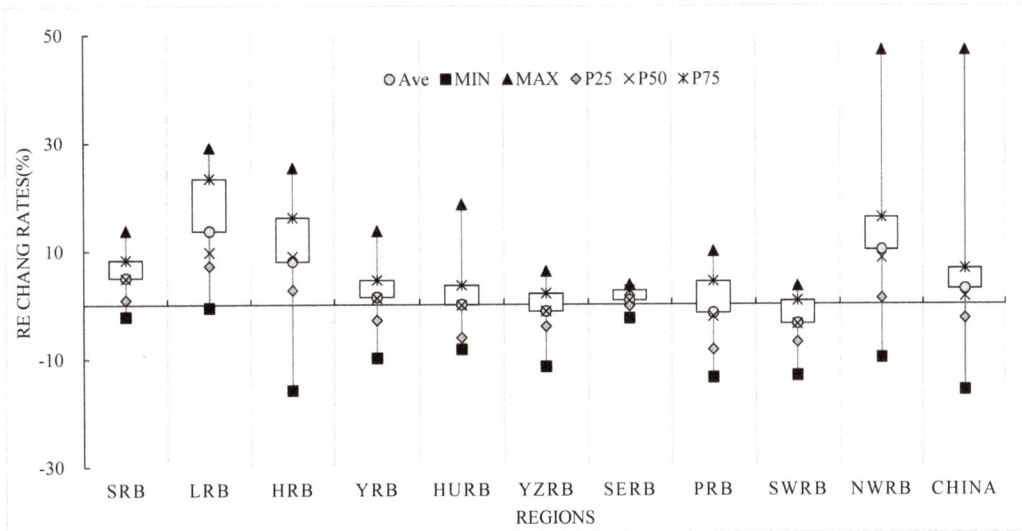

FIGURE 7.13 Box plot of the distribution of relative rates of change of water resources in Class III WRRs under the RCP4.5 scenario.

REFERENCES

Alcamo, J., Döll, P., Henrichs, T., et al., 2003. Development and testing of the WaterGAP 2 global model of water use and availability. Hydrological sciences journal, 48(3), 317–337.

Bai, P., Liu, X. and Liu, C., 2018. Improving hydrological simulations by incorporating GRACE data for model calibration. Journal of hydrology, 557, 291–304.

Donnelly, C., Andersson, J.C.M. and Arheimer, B., 2016. Using flow signatures and catchment similarities to evaluate the e-HYPE multi-basin model across Europe. Hydrological sciences journal, 61(2), 255–273.

FAO (Food and Agriculture Organization), 2003. Review of the world water resources by country. Water Report No. 23, Rome.

Feng, W., Shum, C.K., Zhong, M., et al., 2018. Groundwater storage changes in China from satellite gravity: An overview. Remote sensing, 10(5), 674.

Gao, H., Wang, J., Yang, Y., et al., 2021. Permafrost hydrology of the Qinghai-Tibet Plateau: A review of processes and modeling. Frontiers in earth science, 8, 576838.

Gleeson, T., Wada, Y., Bierkens, M.F., et al., 2012. Water balance of global aquifers revealed by groundwater footprint. Nature, 488, 197–200.

Jia, Y., Wang, H., and Zhou, Z. et al., 2006. Development of the WEP-L distributed hydrological model and dynamic assessment of water resources in the Yellow River basin. Journal of hydrology, 331(3-4), 606–629.

Kirkham, M., 2014, Leaf Anatomy and Leaf Elasticity. In Principles of Soil and Plant Water Relations, 2nd ed. Academic Press, Boston, MA, USA, 409–430.

Kumar, R., Livneh, B. and Samaniego, L., 2013. Toward computationally efficient large-scale hydrologic predictions with a multiscale regionalization scheme. Water resources research, 49(9), 5700–5714.

Lei, Y., Yao, T., Yang, K., et al., 2017. Lake seasonality across the Tibetan Plateau and their varying relationship with regional mass changes and local hydrology. Geophysical research letters, 44(2), 892–900.

Liang, X. and Xie, Z., 2001. A new surface runoff parameterization with subgrid-scale soil heterogeneity for land surface models. Advances in water resources, 24(9–10), 1173–1193.

Luo, Y., Wang, X., Piao, S., et al., 2018. Contrasting streamflow regimes induced by melting glaciers across the Tien Shan–Pamir–North Karakoram. Scientific reports, 8(1), 1–9.

McVicar, T.R., Roderick, M.L., Donohue, R.J., et al., 2012. Global review and synthesis of trends in observed terrestrial near-surface wind speeds: Implications for evaporation. Journal of hydrology, 416, 182–205.

Miao, Y. and Wang, A., 2020. A daily 0.25°× 0.25° hydrologically based land surface flux dataset for conterminous China, 1961–2017. Journal of hydrology, 590, 125413.

Schuol, J., Abbaspour, K.C., Yang, H., et al., 2008. Modeling blue and green water availability in Africa. Water resources research, 44(7), W07406.

Sun, S., Bi, Z., Zhou, S., et al., 2021. Spatiotemporal shifts in key hydrological variables and dominant factors over China. Hydrological processes, 35(8), 14319.

Widén-Nilsson, E., Halldin, S. and Xu, C., 2007. Global water-balance modelling with WASMOD-M: Parameter estimation and regionalisation. Journal of hydrology, 340(1–2), 105–118.

CHAPTER 8

Detection and Attribution of Streamflow Changes in Different Climatic and Geomorphic Regions of China

8.1 RESEARCH BACKGROUND

According to the water resources assessment, it can be known that surface water resources are the main body of China's water resources, accounting for 93.5% of the total, while streamflow is the ultimate expression of surface water resources. Taking advantage of the China Water Assessment Model's (CWAM's) strengths in describing the complex runoff generation mechanism in China, the study further selected 21 typical river basins to quantify the differences in streamflow changes in different climatic and geomorphic regions. Besides, the attribution analysis of streamflow changes in these river basins was conducted. Through large-scale hydrological modeling, understanding the impact of different climatic and geomorphic conditions on streamflow is valuable for water and land management.

Currently, the analysis of streamflow changes and their attribution has become a hot area. In general, two main approaches, model-based and statistics-based methods, are applied for attribution analysis of streamflow changes. The model-based method commonly relies on the distributed physically based hydrological model, where the behavior of model parameters accords with the physical mechanism of the hydrological process, and thus, the hydrological heterogeneity is fully considered. Consequently, the method has greater advantages than the latter in quantifying hydrological responses to climate change and human activities, especially for future scenarios. Consequently, the method is a powerful supplement to the traditional statistics-based method, which is mainly based on the Budyko hypothesis and climate elasticity (Budyko, 1974). Duethmann et al. (2015) assessed the individual contributions to streamflow changes in snow and glacier melt-dominated

DOI: 10.1201/9781003646648-8

catchments using the WASA (Water Availability in Semi-Arid Environments) model and multiple linear regression. As concluded by the authors, "Comparing the two approaches, an advantage of the simulation-based approach is the fact that it is based on process-based relationships implemented in the hydrological model instead of statistical links in the regression model". Zhao et al. (2013) investigated the streamflow increase in the Aksu catchment by applying the VIC (variable infiltration capacity) model, and they attributed the streamflow increase mostly to an increase in precipitation. Setti et al. (2020) applied the SWAT (soil water assessment tool) model to separate the impact of land use and rainfall variability on streamflow in a tropical river basin, India. The study pointed out that the model-based approach provides a better alternative way to understand the impact of the changes on the hydrology of the system. Due to the complexity of the model structure and parameters, most of the available studies focus on a river basin (van Roosmalen et al., 2007; Xu et al., 2013; Eregno et al., 2013). Moreover, many models only consider human activities such as land cover change, without considering the impact of human water consumption. However, there is a lack of research on the quantitative differentiation of detection and attribution of streamflow changes in multiple basins that have different climatic and underlying surface conditions, especially across a vast country like China. According to Huo et al. (2021), the effects of climatic and geographical factors on the long-term water balance in different climatic zones need to be further quantified.

In large-scale regions, the spatial heterogeneity of climatic characteristics indicates different hydrothermal conditions and vegetation cover. Thus, this spatial heterogeneity is the fundamental driving force of the difference in the water balance of ecosystems and the associated streamflow. Furthermore, focusing on the basins located in the same climatic zone, there may be several differences in geomorphic structure that lead to diverse precipitation redistribution and streamflow responses. For example, the hydrological behaviors of karst landforms are very special compared with those of non-karst areas. Additionally, human activities exhibit remarkable regional variability, which significantly affects land use and vegetation patterns as well as river flow regimes. Hence, if we are committed to serving the large-scale water resources planning and management, a deep understanding and quantification of regional differences in streamflow changes is necessary and urgent.

China covers a large land area of about 9.6 million km^2, encompassing many river basins and various geophysical and climatic zones. As the world's second-largest economy, the country has been facing a severe water crisis. Most river basins in north China, such as the Haihe River Basin, the Yellow River Basin, and the Shiyang River Basin, are obviously decreasing in annual streamflow. In the southern streamflow-rich regions, extreme flood or drought events are expected to intensify. To meet the practical needs of nationwide water resources planning and management, we have developed the CWAM by considering different climatic and hydrological conditions, geomorphic structures, and their effects on streamflow. Based on the model, the objective of this chapter was to clarify the differences in the variation characteristics of streamflow from different climatic and geomorphic regions nationwide by trend detection as well as attribution analysis. Therefore, 21 representative river basins in 9 climatic zones and 4 geomorphic regions were selected as study areas. Furthermore, a comparative analysis was conducted on the evolution of streamflow in these river basins from 1956 to 2017.

8.2 DATA AND METHODS

8.2.1 Study Area

China covers 10 Climatic Zones, and the change of these climatic zones from north to south forms roughly three bands, as shown in Figure 8.1. Specifically, the band I is Frigid Temperate Zone (FTZ) – Median Temperate Zone (MTZ) – Warm Temperate Zone (WTZ) – North Asian Tropical Zone (NATZ) – Middle Asian Tropical Zone (MATZ) – South Asian Tropical Zone (SATZ) – Marginal Tropical Zone (MTPZ); the band II is Median Temperate Zone (MTZ) – Alpine Sub-Frigid Zone (ASZ) – Plateau Temperate Zone (PTZ) – Middle Asian Tropical Zone (MATZ) – South Asian Tropical Zone (SATZ); the band III is Median Temperate Zone (MTZ) – Warm Temperate Zone (WTZ) – Alpine Frigid Zone (AFZ) – Alpine Sub-Frigid Zone (ASZ) – Plateau Temperate Zone (PTZ).

Moreover, it is a mountainous country, with 69% of its land comprising mountains, hills, and highlands. Influenced by diverse climates and landforms, China's geomorphic structure is complex. Roughly speaking, four typical geomorphic structures are widely distributed in the country: ① The mountainous area in northern China is characterized

FIGURE 8.1 Distribution of 21 representative river basins, 10 Climatic Zones, and their t3 change bands.

by the soil-bedrock structure. At present, the traditional hydrological modeling is mostly applied to this geomorphic structure; ② The karst mountain region of southwest China (KMRSC) is one of the largest continuous karst areas in the world, which owns thin surface soil, high soil infiltration capacity, and complex topography; ③ The swelling soils, special soil type distributed widely all around the Loess Plateau. The deformable soil will swell, shrink, or collapse along with the movement of soil water; ④ While in vast cold regions, the soil alternately freezes and melts as its temperature decreases and increases, which influences water evaporation and infiltration.

To clarify the spatial heterogeneity of streamflow changes in different climatic and geomorphic regions, 21 representative river basins and their outlet hydrological stations were selected nationwide in China. These river basins involve nine climatic zones, except AFZ, which is located on the Qinghai-Tibet Plateau and lacks river systems. Figure 9.1 illustrates the location of the representative river basins and 10 climatic zones. Meanwhile, all four geomorphic structures were covered. The 21 river basins mostly cover an area of 10,000–100,000 km² and are well represented in terms of climatic and geomorphic characteristics. The basic information of these basins is summarized in Table 8.1.

8.2.2 Data Preparation

This chapter focused on quantifying regional differences in the evolution characteristics of streamflow in 21 representative river basins through the CWAM. In this chapter, some of the key variables were analyzed, including precipitation, actual evapotranspiration, and streamflow. Precipitation data were obtained by spatial interpolation from meteorological stations across the country, while actual evapotranspiration and streamflow were simulated by the model. The results were performed at monthly and annual scales from 1956 to 2017. Additionally, the 1km grid land use data of six periods (1980, 1990, 1995, 2000, 2005, 2010, and 2015) were processed to identify the impact of land use change on streamflow. These data were from the Chinese Academy of Sciences (http://www.dsac.cn/DataProduct). Climatic Zones and landform distribution were available at the Geospatial Data Cloud website of the Computer Network Information Center of the Chinese Academy of Sciences (http://www.gscloud.cn). Soil data, including a map of the 1 km grid soil types and the corresponding characteristic parameters, were obtained from the National Second Soil Survey Data. Streamflow data were from the National Second Water Resources Survey Data.

8.2.3 Methodology
8.2.3.1 Analysis of Intra-Annual Concentration of Hydrological Variables

Concentration period (Cp) and Concentration degree (Cd), proposed by Zhang and Qian (2003), are two important indexes to measure the annual concentration of hydrological variables. The Cp represents the period (month) in which the total amount within a year concentrates, and the Cd represents the degree to which the total amount is distributed in 12 months (Yeşilırmak and Atatanır, 2016). The basic idea for calculating Cp and Cd is that the monthly value of hydrological variables is a vector quantity having both magnitude and direction. The directions are the angles assigned to each month in 30° increments, all

TABLE 8.1   Basic Information of 21 Representative River Basins across China

River basin (Area/10⁴ km²)	Hydrological Station	Location	Climatic Characteristics	Geomorphic Structures
Jiliuhe River (0.77)	Mangui	E122.05°N52.05°	Frigid Temperate Zone; Humid climate	Low mountain, hills; freezing and melting soils
Nenjiang River (5.71)	Ayanqian	E124.63°N48.77°	Median Temperate Zone; Semi-humid climate	Low mountain, hills; freezing and melting soils
Hulan River (2.77)	Lanxi	E126.33°N46.25°	Median Temperate Zone; Semi-humid climate	Hills, plain; freezing, and melting soils
Hunjiang River (1.49)	Shajianzi	E125.43°N41.00°	Median Temperate Zone; Humid climate	Middle and low mountains, hills; soil-bedrock structure
Jimulun River (1.83)	Meilinmiao	E120.90°N43.98°	Median Temperate Zone; Semi-arid climate	Middle and low mountains, hills; soil-bedrock structure
Luanhe River (4.41)	Luanxian	E118.76°N39.74°	Warm Temperate Zone; Semi-humid and semi-arid climate	Middle and low mountains; soil-bedrock structure
Zhanghe River (1.78)	Guantai	E114.08°N36.33°	Warm Temperate Zone; Semi-humid and semi-arid Climate	Middle and low mountains; soil-bedrock structure
Weihe River (4.68)	Xianyang	E108.69°N34.32°	Warm Temperate Zone; Semi-humid climate	Low mountains, hills; swelling soil in Loess Plateau
Yihe River (1.03)	Linyi	E118.40°N35.02°	Warm Temperate Zone; Transition zone of semi-humid and Semi-arid climate	Middle and low mountains; soil-bedrock structure
Huaihe River (1.60)	Huaibin	E115.41°N32.44°	North Asian Tropical Zone; Humid and semi-humid climate	Hills, plain; soil-bedrock structure
Xiangjiang River (8.16)	Xiangtan	E112.92°N27.86°	Middle Asian tropical Zone; Humid climate	Middle & low mountains; soil-bedrock structure
Minjiang River (5.45)	Zhuqi	E119.10°N26.65°	Middle Asian tropical Zone; Humid climate	Middle & low mountains; soil-bedrock structure
Wujiang River (5.13)	Sinan	E108.25°N27.94°	Middle Asian tropical Zone, Humid climate	Karst mountain region of southwest China
Longjiang River (1.58)	Sancha	E108.95°N24.47°	South Asian Tropical Zone; Humid climate	Karst mountain region of southwest China
Yujiang River (7.27)	Nanning	E108.23°N22.83°	South Asian Tropical Zone; Humid climate	High and middle mountains; soil-bedrock structure
Dongjiang River (2.53)	Boluo	E114.30°N23.17°	South Asian Tropical Zone; Humid climate	Middle and low mountains; soil-bedrock structure
Nandu River (0.68)	Longtang	E110.42° 19.88°	Marginal Tropical Zone; Humid climate	High and middle mountains; soil-bedrock structure
Yellow River (12.20)	Tangnaihai	E100.15°N35.50°	Alpine Sub-frigid Zone; Semi-arid and semi-humid climate	High mountains; freezing and melting soils in Tibetan Plateau
Heihe River (1.00)	Yingluoxia	E100.20°N38.82°	Alpine Sub-frigid Zone; Semi-arid and arid climate	High mountains; freezing and melting soils
Yalongjiang River (11.70)	Xiaodeshi	E101.83°N26.76°	Plateau Temperate Zone and Middle Asian tropical Zone; Humid climate	High mountains; Freezing and melting soils
Yalu Zangbu River (10.61)	Nugesha	E89.71° N29.32°	Plateau Temperate Zone; Semi-arid climate	High and middle mountains; freezing and melting soils in Tibetan Plateau

TABLE 8.2   Corresponding Relation between Month, Cp, and Cpm

Month	Jan.	Feb.	Mar.	Apr.	May	Jun.	Jul.	Aug.	Sep.	Oct.	Nov.	Dec.
Cp	0°	30°	60°	90°	120°	150°	180°	210°	240°	270°	300°	330°
Cp_m	1.5	2.5	3.5	4.5	5.5	6.5	7.5	8.5	9.5	10.5	11.5	12.5

of which comprise a circle (360°) for a year (Table 8.2). The procedure to calculate Cp and Cd includes equations (8.1)–(8.5):

$$R_i = \sum_{j=1}^{12} r_{ij} \tag{8.1}$$

$$R_{xi} = \sum_{j=1}^{12} r_{ij} \cdot \cos\theta_j \tag{8.2}$$

$$R_{yi} = \sum_{j=1}^{12} r_{ij} \cdot \sin\theta_j \tag{8.3}$$

$$Cp_i = \arctan(R_{xi}/R_{yi}) \tag{8.4}$$

$$Cd_i = \sqrt{R_{xi}^2 + R_{yi}^2}/R_i \tag{8.5}$$

where i is the year; j is the month in a year; r_{ij} is the value of the hydrological variable in the jth month of the ith year; θi is the angle assigned to the month j, –.

According to the above calculation results, the Cp is a value of 0–360. To represent the month more intuitively in which hydrological variables are concentrated, this chapter uses linear interpolation to convert Cp to the value of 1–13. The formula is equation (8.6):

$$Cp_m = \begin{cases} \dfrac{1}{30} \cdot Cp + 1.5 & 0 \le Cp \le 345 \\[2ex] \dfrac{1}{30} \cdot (Cp - 360) + 1.5 & 345 < Cp \le 360 \end{cases} \tag{8.6}$$

where the integer part of the Cp_m represents the month, and the decimal part reflects the position in this month. The larger the decimal part is, the more concentrated the hydrological variables are at the end of the month. For example, the 1.5 in Table 8.2 represents that the hydrological variables concentrate in the middle of January.

For these 21 representative river basins, the Cp_m and Cd from 1956 to 2017 were calculated and then used to depict their spatial-temporal changes.

8.2.3.2 Model-based Attribution Analysis of Streamflow Change

Change in streamflow can be attributed to the impact of climate change and human activities, and the human activities mainly involve land use change and water intake. Because the WEP-CN model simulates the process of rainfall-runoff generation without consideration of water use, the total change in river streamflow (ΔQ) was tentatively assumed to be a response to climate change (ΔQ_{cc}) and land use change (ΔQ_{lu}):

$$\Delta Q = \Delta Q_{cc} + \Delta Q_{lu} \tag{8.7}$$

Based on the verified model, the scenario analysis is performed to separate the contribution of climate and land use changes to streamflow variability. First, the time when abrupt changes of streamflow occurred is identified using the M–K test and t-test. Accordingly, the base period (before the abrupt change) and the effect period (after the abrupt change) of the streamflow series are divided. Second, three scenarios are set as the input data of the verified model. Scenario 1: Climatic conditions and land use in the base period; Scenario 2: Climatic conditions in the effect period and land use in the base period; Scenario 3: Climatic conditions and land use in the effect period. Third, with the same model structure and parameters, the mean annual streamflow under scenarios 1, 2, and 3 is simulated, respectively, namely, R_{N1}, R_{C2}, and R_{N2}. The contribution rates of climatic conditions (η_{cc}) and land use change (η_{lu}) on streamflow are finally obtained using equations (8.8) and (8.9):

$$\eta_{cc} = \frac{|\Delta Q_{cc}|}{|\Delta Q_{cc}| + |\Delta Q_{lu}|} \times 100\% \tag{8.8}$$

$$\eta_{lu} = \frac{|\Delta Q_{lu}|}{|\Delta Q_{cc}| + |\Delta Q_{lu}|} \times 100\% \tag{8.9}$$

where $\Delta Q_{cc} = R_{C2} - R_{N1}$, and $\Delta Q_{lu} = R_{N2} - R_{C2}$.

8.3 STREAMFLOW AND ITS COMPOSITION CHANGES

8.3.1 Validation of the CWAM in 21 Representative River Basins

The simulated streamflows were used to compare with the naturalized streamflows by reverting statistical water consumption to the observed ones. The relative error (RE) of mean annual streamflow and the Nash-Sutcliffe efficiency (NSE) of monthly streamflow in each basin were calculated, as shown in Table 8.3. For the calibration period

TABLE 8.3 Calibration and Validation of the CWAM in 21 Representative River Basins

Hydrological Station	Calibration Period		Validation Period		Hydrological Station	Calibration Period		Validation Period	
	NSE	RE	NSE	RE		NSE	RE	NSE	RE
Mangui	0.64	−5.0%	0.65	4.3%	Xiaodeshi	0.94	−3.9%	0.92	3.0%
Ayanqian	0.67	−2.8%	0.72	1.3%	Xiangtan	0.94	3.3%	0.90	8.3%
Lanxi	0.76	7.9%	0.72	0.9%	Sinan	0.86	8.8%	0.88	4.1%
Shajianzi	0.73	1.5%	0.83	−2.9%	Sancha	0.88	2.0%	0.91	−1.8%
Meilinmiao	0.64	4.1%	0.79	−8.2%	Nanning	0.91	0.8%	0.90	−3.0%
Luanxian	0.78	−2.3%	0.69	−1.8%	Boluo	0.75	−3.8%	0.79	−3.4%
Guantai	0.80	−2.9%	0.81	5.1%	Longtang	0.79	5.4%	0.80	−2.9%
Huaibin	0.90	−6.8%	0.92	3.3%	Zhuqi	0.84	−1.2%	0.82	−2.3%
Linyi	0.76	0.7%	0.86	1.1%	Nugesha	0.74	−9.9%	0.85	6.0%
Tangnaihai	0.83	−3.9%	0.81	−5.3%	Yingluoxia	0.84	8.0%	0.82	12.5%
Xianyang	0.66	−7.9%	0.69	2.1%					

(1956–1980), the absolute values of *RE* were smaller than 10%, and the *NSE* were larger than 0.64. For the validation period (1981–2000), the absolute values of *RE* were all smaller than 12.5%, and the *NSE* were larger than 0.65. Especially, 97% of the basins own the absolute values of *RE* less than 10%, and 60% of them had the *NSE* value larger than 0.8. According to Moriasi et al. (2007), $NSE > 0.50$ can be considered a satisfactory simulation. Overall, it can be concluded that WEP-CN performs well and is applicable to the study basins.

8.3.2 Characteristics of Changes in Precipitation and Actual Evapotranspiration

8.3.2.1 Precipitation

Table 8.4 illustrates the annual value of precipitation and its concentration degree within the year in 21 representative river basins. The precipitation evolution varied greatly among different climatic and geomorphic regions throughout China. The mean annual precipitation (*MAP*) increased gradually along the band I (FTZ – MTZ – WTZ – NATZ – MATZ – SATZ – MTPZ), varying between 538 and 2339 mm. Meanwhile, there existed significant

TABLE 8.4 Temporal and Spatial Variation of Annual Precipitation, PCD, and PCP in the 21 Representative River Basins

Climatic Zones	River Basins	Hydrological Stations	MAP/mm	PCD	PCP	Z Values of M-K test		
						Annual Precipitation	PCD	PCP
Frigid Temperate Zone (FTZ)	Jiliuhe River	Mangui	538	0.68	7.82	0.69	−3.53**	−4.13**
Median Temperate Zone (MTZ)	Nenjiang River	Ayanqian	561	0.70	7.82	−0.66	−2.30*	−4.12**
	Hulan River	Lanxi	611	0.69	7.76	−1.14	−4.44**	−3.88**
	Hunjiang River	Shajianzi	957	0.62	7.80	−1.32	−4.62**	−2.86**
	Jimulun River	Meilinmiao	408	0.68	7.82	−1.25	−4.88**	−3.24**
Warm Temperate Zone (WTZ)	Luanhe River	Luanxian	563	0.71	7.73	−0.93	−4.47**	−3.05**
	Zhanghe River	Guantai	590	0.65	7.87	−1.56	−2.81**	−1.89
	Weihe River	Xianyang	562	0.53	7.78	−1.68	0.93	−0.98
	Yihe River	Linyi	864	0.63	7.77	−1.24	−1.19	−0.37
North Asian Tropical Zone (NATZ)	Huaihe River	Huaibin	1175	0.43	7.29	−1.59	−0.54	−0.88
Middle Asian tropical Zone (MATZ)	Xiangjiang River	Xiangtan	1672	0.32	5.91	0.00	0.45	−0.18
	Minjiang River	Zhuqi	1943	0.37	5.92	−0.26	−0.11	−0.69
	Wujiang River	Sinan	1293	0.44	7.41	−3.02**	0.65	−1.72
South Asian Tropical Zone (SATZ)	Longjiang River	Sancha	1650	0.44	6.98	−0.73	0.25	−2.68**
	Yujiang River	Nanning	1523	0.49	7.50	−1.56	0.61	−1.10
	Dongjiang River	Boluo	2034	0.45	6.55	−1.69	1.29	−1.66
Marginal Tropical Zone (MTPZ)	Nandu River	Longtang	2339	0.44	8.40	−1.32	1.40	0.18
Alpine Sub-frigid Zone (ASZ)	Yellow River	Tangnaihai	515	0.61	7.60	2.30*	−1.66	−2.52*
	Heihe River	Yingluoxia	375	0.72	7.76	5.43**	−5.43**	−3.12**
Plateau Temperate Zone (PTZ)	Yalongjiang River	Xiaodeshi	814	0.64	7.83	0.18	−2.01*	−4.18**
	Yalu Zangbu River	Nugesha	378	0.70	8.01	1.17	−4.30**	−4.16**

differences in annual precipitation between the two basins located in the same climatic zone. For example, in the four watersheds located in the MTZ, the maximum and minimum *MAP* values were 957 mm and 408 mm, respectively. Interestingly, the precipitation concentration degree (*PCD*) in 21 basins showed a distribution pattern opposite to that of precipitation. The value was generally larger than 0.7 in arid or semi-arid basins, but less than 0.5 in southern humid basins. The results showed that compared with humid areas in China, precipitation distribution in arid and semi-arid areas was more concentrated, which may make water resource utilization more difficult. Moreover, the *PCD* was strongly positively correlated with latitude at a 0.01 confidence level. Compared with the southern humid basins, the precipitation in the arid basins is less but more concentrated. Consequently, the effective utilization of water resources in the water-deficient basins in northern China faces greater challenges. In terms of the precipitation concentration period (*PCP*), the northern basins were about mid-July to early August, which was generally one month later than that in the southern basins.

The *MAP* in most basins exhibited a downward trend, whereas significant increases were detected in several basins that are located in the Qinghai-Tibet Plateau and its surrounding areas. In particular, the precipitation in the source areas of the Yellow River and Heihe River increased most significantly, while their concentration degree decreased, and the concentration period advanced. This trend may be beneficial to the streamflow recharge and utilization of the downstream basins. Focusing on the change trend of the *PCD*, different climatic zones had a distinct regional pattern. Specifically, the *PCD* in the southern tropical basins tended to increase. Although the trend was not significant, it is still necessary to pay more attention to the possible increase of flood pressure in these basins. On the contrary, the *PCD* in most of the northern temperate and cold basins (i.e., the MTZ, ASZ, and PTZ) showed a significant decreasing trend, and the distribution of precipitation within a year tended to be uniform. Besides, almost all basins had an earlier precipitation concentration period than in the past, except the Nandu River Basin (above Longtang), and the trend was notable in northern temperate and cold basins.

8.3.2.2 *Actual Evapotranspiration*

To reflect the dry-wet conditions of the basins, in addition to calculating the actual evapotranspiration (*AET*) of each representative river basin, the proportion of annual actual evapotranspiration in precipitation (*AET/P*) was also analyzed. The results are shown in Table 8.5.

Mean annual actual evapotranspiration (*AETm*) increased gradually from cold to tropical basins, ranging between 272 and 1753 mm. There existed a strong consistency between *MAP* and *AETm* across these basins, as the Pearson correlation coefficient of the two reached 0.95, which passed the confidence level of 0.01. The value of *AET/P* showed significant climatic zone differences, and increased first and then decreased along the band I (FTZ – MTZ – WTZ – NATZ – MATZ – SATZ – MTPZ). The high values were mainly distributed in MTZ and WTZ, where more than 80% of the precipitation was consumed by the evapotranspiration. Affected by the low air temperature in FTZ and ASZ, the

TABLE 8.5 Temporal and Spatial Variation of Actual Evapotranspiration and Its Proportion in Precipitation in the 21 Representative River Basins

Climatic Zones	River Basins	Hydrological Stations	AETm/mm	MeanAnnual AET/P	Z Values of M-K Test AETm	AET/P
Frigid Temperate Zone (FTZ)	Jiliuhe River	Mangui	310.81	0.59	−0.33	−1.43
Median Temperate Zone (MTZ)	Nenjiang River	Ayanqian	409.35	0.75	−1.31	−0.84
	Hulan River	Lanxi	470.05	0.77	−1.46	−0.25
	Hunjiang River	Shajianzi	525.79	0.56	−2.75**	0.62
	Jimulun River	Meilinmiao	398.23	1.00	−1.45	−0.2
Warm Temperate Zone (WTZ)	Luanhe River	Luanxian	490.61	0.89	−0.03	1.26
	Zhanghe River	Guantai	525.12	0.92	−2.20**	0.83
	Weihe River	Xianyang	459.47	0.83	4.14**	2.79**
	Yihe River	Linyi	621.34	0.75	−3.73**	−0.08
North Asian Tropical Zone (NATZ)	Huaihe River	Huaibin	818.75	0.72	−3.88**	0.28
Middle Asian tropical Zone (MATZ)	Xiangjiang River	Xiangtan	812.20	0.5	−5.32**	−1.54
	Minjiang River	Zhuqi	1076.01	0.57	−1.05	−0.29
	Wujiang River	Sinan	733.67	0.58	−3.13**	0.42
South Asian Tropical Zone (SATZ)	Longjiang River	Sancha	874.99	0.54	−0.88	0.16
	Yujiang River	Nanning	936.26	0.63	−1.59	1.19
	Dongjiang River	Boluo	1223.03	0.62	1.04	1.97*
Marginal Tropical Zone (MTPZ)	Nandu River	Longtang	1753.28	0.78	1.01	1.78
Alpine Sub-frigid Zone (ASZ)	Yellow River	Tangnaihai	332.79	0.65	5.19**	2.57*
	Heihe River	Yingluoxia	271.87	0.74	5.10**	−4.50**
Plateau Temperate Zone (PTZ)	Yalongjiang River	Xiaodeshi	393.25	0.49	0.13	0.11
	Yalu Zangbu River	Nugesha	272.51	0.73	1.43	−0.57

evapotranspiration was limited. Although the precipitation was small, the AET/P had a low value. Furthermore, the $AETm$ was large in the tropical zones (i.e., NATZ, MATZ, SATZ, MTPZ), but it only accounted for 0.5–0.7 of the local precipitation.

The AET of most basins was decreasing, covering FTZ, MTZ, WTZ, NATZ, MATZ, and SATZ. Among them, the trend was most significant in the Xiangjiang River basin (above Xiangtan), the Huaihe River basin (above Huaibin), the Wujiang River basin (above Sinan), and the Yihe River basin (above Linyi). However, the Weihe River basin (above Xianyang) showed a significantly increased trend opposite to that of the surrounding basins. The main reason is that China has implemented large-scale afforestation program on the Loess Plateau since 1999 (Wang et al., 2016). Other basins where the AET had an increased trend were mainly concentrated in the Qinghai-Tibet Plateau and its surrounding areas. Affected by the increasing precipitation and air temperature in ASZ, the source area of Heihe River Basin (above Yingluoxia) and the Yellow River (above Tangnaihai) exhibited a significant increasing trend in the AET. For different basins in the same climatic zone, the trend of the AET/P was less consistent than that of the AET. Obviously, the impact of the change of climatic conditions and land use on the AET/P was more complex than that on the AET.

It should be noted that the source area of the Yellow River Basin (above Tangnaihai) and the Weihe River Basin (above Xianyang), located in the Loess Plateau, were evolving toward dryness. The notable increase of the *AET/P* means that more precipitation translates into evapotranspiration rather than streamflow, which has led to the decrease of water storage in these basins. The reason for this phenomenon may be related to the local excessive afforestation, but the reasonable scale of forest and grass planting needs to be further studied. Some researchers have urged a cessation of the Grain for Green expansion on the Loess Plateau in China because it leads to an annual decline in the streamflow of the Yellow River (Feng et al., 2016). Although the *AET* increased in the source area of the Heihe River Basin (above Yingluoxia), its proportion in precipitation decreased significantly. This showed that the basin has been under climatic warming and wetting conditions in recent decades. However, due to the intensification of limited glacial meltwater, whether this phenomenon is a long-term trend needs to be viewed more carefully. Moreover, the basins in WTZ were characterized by decreased precipitation and increased *AET/P* values, which further confirmed the phenomenon of warming and drying in the Huabei region of north China (Ma and Fu, 2003). The basins in the cold region of northeast China were experiencing a more humid climate, while the southern basins tended to be drier, although these trends were not significant.

8.3.3 Characteristics of Changes in Streamflow and Its Components

In this chapter, the soil flow and underground flow were classified as river baseflow, and then the baseflow index (BFI) was calculated as the proportion of river baseflow in streamflow. Consequently, the temporal and spatial variation of streamflow and its components in 21 representative river basins were analyzed (Table 8.6).

As expected, the spatial distribution of annual streamflow was highly consistent with that of precipitation, as their Pearson correlation coefficient reached 0.96. However, the trend characteristics of streamflow were different from those of precipitation and actual evapotranspiration. Among the seven basins affected by the freezing and melting soils, the annual runoff depth of five basins showed an increasing trend. The reason may be that increasing air temperature leads to the melting of ice and snow and frozen soil in cold regions, which increases the contribution to streamflow. In addition, the Weihe River Basin (above Xianyang), Luanhe River Basin (above Luanxian), and Zhanghe River Basin (above Guantai) were the basins with the most significant decrease in natural streamflow. These basins are concentrated in the Warm Temperate Zone with intense human activities and a fragile ecosystem. The reduction of streamflow will further aggravate local water scarcity.

Furthermore, the spatial differences of BFI among the 21 river basins were analyzed. On the whole, the proportion of river baseflow in streamflow in the northern basins was higher than that in the southern basins. Except for the Nenjiang River Basin (above Ayanqian), Jiliuhe River Basin (above Mangui), and Hunjiang River Basin (above Shajianzi), the BFI of other basins all exceed 0.5. Especially in the Heihe River Basin (above Yingluoxia), the value reached 0.78. The BFI in the southern basins was generally below 0.5, except for the Nandu River basin (above Longtang). The reason may be that

TABLE 8.6   Temporal and Spatial Variation of Streamflow and Its Baseflow Index in the 21 Representative River Basins

Climatic Zones	River Basins	Hydrological Stations	Annual Streamflow Depth/mm	Baseflow Index	Z values of M-K Test Runoff Depth	AET/P
Frigid Temperate Zone (FTZ)	Jiliuhe River	Mangui	199.35	0.43	0.98	−0.98
Median Temperate Zone (MTZ)	Nenjiang River	Ayanqian	159.07	0.35	0.83	−0.28
	Hulan River	Lanxi	136.79	0.57	1.04	0.83
	Hunjiang River	Shajianzi	428.59	0.50	−0.98	1.51
	Jimulun River	Meilinmiao	12.02	0.56	1.64	−0.88
Warm Temperate Zone (WTZ)	Luanhe River	Luanxian	73.11	0.67	−1.98*	−0.62
	Zhanghe River	Guantai	77.13	0.69	−2.93**	−0.83
	Weihe River	Xianyang	110.04	0.61	−4.06**	1.59
	Yihe River	Linyi	287.67	0.51	−1.02	0.46
North Asian Tropical Zone (NATZ)	Huaihe River	Huaibin	357.25	0.26	1.52	0.22
Middle Asian tropical Zone (MATZ)	Xiangjiang River	Xiangtan	914.94	0.41	0.8	−1.39
	Minjiang River	Zhuqi	934.29	0.39	−0.05	−0.88
	Wujiang River	Sinan	557.33	0.48	−1.69	0.87
South Asian Tropical Zone (SATZ)	Longjiang River	Sancha	787.78	0.49	−0.42	0.35
	Yujiang River	Nanning	527.08	0.41	−1.31	−0.2
	Dongjiang River	Boluo	864.35	0.37	−0.98	1.45
Marginal Tropical Zone (MTPZ)	Nandu River	Longtang	962.79	0.60	−1.29	−0.84
Alpine Sub–frigid Zone (ASZ)	Yellow River	Tangnaihai	163.35	0.59	−0.63	2.55*
	Heihe River	Yingluoxia	170.50	0.78	1.66	3.17**
Plateau Temperate Zone (PTZ)	Yalongjiang River	Xiaodeshi	428.02	0.39	0.26	−0.04
	Yalu Zangbu River	Nugesha	134.62	0.65	−0.76	−0.20

Note: superscripts * and ** denote significance tests that pass the confidence levels of 0.05 and 0.01, respectively. Same as below.

the precipitation in the northern basins from October to March is less than that in the southern basins, and river streamflow mainly relies on groundwater. However, when the climatic conditions are similar, the difference in geomorphic structures becomes an important factor in changing the streamflow components of the basin. In the Longjiang River Basin (above Sancha) and Wujiang River Basin (above Sinan), characterized by karst geomorphic structure, the BFI reached 0.49 and 0.48, respectively. However, for other basins located in the same climatic zone as the two basins, the value was only around 0.4. In the karst mountain region, the epikarst zone with large interconnected aquifers can significantly enhance the water storage capacity of the groundwater aquifer, increasing groundwater-river interaction. According to the trend detection, the BFI of the source areas of the Yellow River and Heihe River increased most significantly. Due to a larger-scale melting of frozen soil, soil infiltration capacity and relative water storage capacity increased. Affected by the ecological restoration program, the streamflow of the Weihe River basin (above Xianyang) reduced significantly while the BFI increased.

This has not been paid much attention to in previous studies, and its possible ecological impacts should be further explored.

8.4 CONTRIBUTIONS OF CLIMATE AND LAND USE CHANGES TO STREAMFLOW

The base and effect periods of natural streamflow in each river basin were determined, and then the contributions of climate and land use changes to streamflow were quantified (Table 8.7). By observing the η_{cc} and η_{lu} values, climate change played a role in reducing streamflow for most basins. Only 6 of the 21 representative river basins experienced an increase in streamflow, which were mainly in the northern cold region. In comparison, the positive or negative contribution of land use change to streamflow was more complex in space. Affected by land use change, the streamflow generally decreased in the WTZ but mainly increased in the ASZ and MTZ. However, it was not clear whether the contribution of land use change to streamflow is positively or negatively dominant in other climatic zones.

The contribution rate of climate and land use changes to streamflow was significantly different among these river basins. Meanwhile, the two possessed opposite contribution characteristics to streamflow change in river basins like the Jiliuhe River basin (above Mangui). For most basins (86% of the total), climate change contributed more to streamflow increase or decrease, as the η_{cc} values varied from 51.4% to 95.7%. Briefly, there existed three high-value areas of η_{cc}, namely the Jiliu River basin in the FTZ, the Dongjiang, Wujiang, and Yujiang River basins in the southwest, and the Qinghai-Tibet Plateau and its surrounding areas. These basins span north and south of China and involve many climatic zones, which further indicates that the impact of climate change on streamflow is significant and widespread in China.

In contrast, only in the Weihe, Huaihe, and Hulan River basins, the variation of land use dominated the change of natural streamflow. Although the number of river basins was small, the impact of land use change on streamflow and its spatial differences were well reflected. The Hulan River basin (above Lanxi) in the Songnen Plain had a significant land use change, which contributed to 66.1% of the streamflow increase. Intensive agricultural practices that peaked in the 1990s have led to large areas of forest, grassland, and wetland being reclaimed for cropland and residential sites. The proportion of cropland reached 10.75% in 2015, an increase of 56.58% compared to 1980. Correspondingly, the area of forests has been greatly reduced, and its proportion decreased from 31.89% to 28.34%. As a result, the capacity of the basin to infiltrate and store precipitation in soil decreased significantly, and more streamflow was generated. Similar results were found in the Huaihe River basin (above Huaibin). However, the land use change in the Weihe River basin (above Xianyang) was opposite to that in the other two basins. The grain to green program (GTGP) has converted 16,000 km² of rain-fed cropland to planted vegetation, causing a 25% increase in vegetation cover during the last decade. As mentioned above, this program has caused a significant increase in evapotranspiration (*AET*) and its proportion in precipitation (*AET/P*). Consequently, land use change contributed 55.5% to the continuous reduction of streamflow.

TABLE 8.7 Contribution Rate and Direction of Climate and Land Use Changes to Streamflow in the Representative Basins

Climatic Zones	Basins	Hydrological Stations	Abrupt Year	ΔQ/mm	η_{cc}/%	Dir_{cc}/-	η_{lu}/%	Dir_{lu}/-
Frigid Temperate Zone (FTZ)	Jiliuhe River	Mangui	1978	18.1	87.2	Increase	12.8	Decrease
Median Temperate Zone (MTZ)	Nenjiang River	Ayanqian	1983	10.5	51.4	Increase	48.6	Increase
	Hulan River	Lanxi	1985	11.6	33.9	Decrease	66.1	Increase
	Hunjiang River	Shajianzi	1971	−27.2	50.3	Decrease	49.7	Decrease
	Jimulun River	Meilinmiao	1984	1.6	56.3	Increase	43.8	Increase
Warm Temperate Zone (WTZ)	Luanhe River	Luanxian	1983	−18.2	69.2	Decrease	30.8	Decrease
	Zhanghe River	Guantai	1977	−22.2	68.0	Decrease	32.0	Decrease
	Weihe River	Xianyang	1992	−26.5	44.5	Decrease	55.5	Decrease
	Yihe River	Linyi	1994	−58.1	59.2	Decrease	40.8	Decrease
North Asian Tropical Zone (NATZ)	Huaihe River	Huaibin	1987	13.0	34.7	Decrease	65.3	Increase
Middle Asian tropical Zone (MATZ)	Xiangjiang River	Xiangtan	1979	84.2	53.1	Increase	46.9	Increase
	Minjiang River	Zhuqi	1984	−34.6	72.0	Decrease	28.0	Decrease
	Wujiang River	Sinan	1996	−66.0	91.0	Decrease	9.0	Increase
South Asian Tropical Zone (SATZ)	Longjiang River	Sancha	1997	−42.1	69.1	Decrease	30.9	Decrease
	Yujiang River	Nanning	1996	−59.0	83.4	Decrease	16.6	Increase
	Dongjiang River	Boluo	1977	−89.8	74.8	Decrease	25.2	Increase
Marginal Tropical Zone (MTPZ)	Nandu River	Longtang	1985	−147.6	69.8	Decrease	30.2	Decrease
Alpine Sub-frigid Zone (ASZ)	Yellow River	Tangnaihai	1986	−3.8	89.6	Decrease	10.4	Increase
	Heihe River	Yingluoxia	2001	23.0	95.7	Increase	4.3	Increase
Plateau Temperate Zone (PTZ)	Yalongjiang River	Xiaodeshi	1979	28.2	78.4	Increase	21.6	Increase
	Yalu Zangbu River	Nugesha	1978	−6.7	95.5	Decrease	4.5	Decrease

Note: Dir_{cc} and Dir_{lu} refer to the directions of climate change and land use contributions to streamflow.

8.5 ANALYSIS AND DISCUSSION OF RESULTS

8.5.1 Reasonable Analysis of Results

It is a challenge to identify regionalized characteristics of streamflow changes in different climatic and geomorphic basins, especially using a physically based, distributed hydrological model. Based on the CWAM, many interesting results were revealed, such as warming and drying in WTZ (i.e., Huabei region), warming and wetting in the Heihe River basin, and streamflow reduction affected by the afforestation program in the Loess Plateau. These results corresponded well with the previous studies. The Haihe River basin, a typical large basin in WTZ, has been one of the most significant areas of warming and drying in China in recent decades. Shi et al. (2007) found that in the Heihe River basin, the annual temperature increased significantly from 1960 to 2009, with an abrupt upward change occurring in about 1986. Precipitation in the high mountains also experienced a substantial increase during this period.

Moreover, some exploratory views were proposed. In total, 21 river basins in 9 climatic zones had an inverse correlation between annual precipitation and its concentration degree, as the Pearson correlation coefficient reached −0.862(p < 0.01). The results quantitatively revealed that, for the northern arid region, the precipitation is less but more concentrated, which brings greater challenges to water resources utilization. By comparing the results of multi-basin analysis, the study elaborated on the impact of karst aquifers and frozen soil on streamflow. For example, karst and frozen soil structure caused a significantly higher BFI than other basins in the same climatic zone. Interestingly, the streamflow decreased, but BFI increased in the Loess Plateau. The annual flattening of streamflow has some benefits to soil and water conservation and ecological improvement in the region. However, some researchers worried that the Grain for Green Expansion would cause a further reduction in the amount of water available to humans.

Due to insufficient existing data, it is difficult to conduct further in-depth and rich research on the interesting phenomena discovered. Through comparative analysis of streamflow characteristics in different river basins, it is found that the frozen soil and karst structures contributed to the increase of the BFI. However, the correlation between the characteristics of frozen soil or karst and the BFI has not yet been quantified. Therefore, we plan to select more basins for further research. First, determine a series of indicators that can scientifically and reasonably characterize permafrost or karst landforms, such as area ratio, etc. Then, through data analysis methods, the quantitative relationship between these indicators and the BFI is established.

8.5.2 Model Uncertainty

The uncertainties from the input data, structures, and parameters of the CWAM, owning nine climatic zones and four geomorphic regions, are the problems that cannot be ignored. In the selection of representative river basins, the study has considered the influence of basin area on streamflow, and the areas of 21 basins have little difference (generally between 10,000 and 100,000 km²). Zhang et al. (2019) established a quantitative relationship between streamflow and basin area in the Haihe River basin, finding that as basin area increased, the average annual streamflow of the basin decreased logarithmically. Huo et al. (2021) found that the average annual streamflow coefficient was strongly correlated with basin area, showing a scale effect. However, this study lacks attention to regional differences of other basin characteristic indicators, such as river length, slope, elevation, etc. Moreover, due to the limitation of data, the number of river basins selected in the comparative study is still relatively small. This may lead to a risk of not fully revealing the impact of special geomorphic structures on streamflow. Meanwhile, it should be noted that as the number of selected basins increases, the uncertainty of model results will also increase.

Overall, the model-based method may underestimate the contribution of land use change (i.e., overestimate the contribution of climate change) to streamflow in a large-scale region. Through a three-dimensional interpolation method considering elevation effects, the meteorological data of the stations have been well distributed to the sub-basins in the CWAM. However, it is difficult to obtain refined land use data for large-scale simulation. Pielke (2005) mentioned that the effects of spatially heterogeneous land use might at least be as important in altering the weather as changes in climate patterns associated with greenhouse gases. As the land use data of 1km × 1km is utilized in this study, the lack of high-resolution data may hide a lot of information about land use change in different periods. This results in the impact of climate change on streamflow in the model being more sensitive to land use change. Taking the Weihe River basin (above Xianyang) as an example, compared with the previous studies (Deng et al., 2020), the conclusion that streamflow decreased significantly in the 1990s and land use change dominated streamflow reduction is consistent. However, the cumulative contribution rate of land use change to streamflow (55.5%) is lower than that in the above studies. Correspondingly, the contribution of climate change is relatively higher in this study. To relieve this potential bias, the model requires refining land use data in simulations by interpreting high-resolution remote sensing data sets. In addition, some scholars attributed conflicting results regarding the effects of land use and climate change on streamflow to uncertainties in hydrological simulations (Yin et al., 2017).

Although the lack of fine land use data and model uncertainties may impact the results, this study can well reveal the combined effects of climatic and underlying surface conditions on hydrological processes and their spatial heterogeneity. Meanwhile, some meaningful conclusions were obtained, which can be used as a guide for practical water and land management. Research has found that most river basins in the northern climate zone of China had a trend of warming and drying, which showed increased stress on water resources in the north. Although the cold regions showed an increase in streamflow, to which the thawing processes made a non-negligible contribution, the cold regions may continue to encounter water shortages as future thawing water decreases. Moreover, there may be a problem of excessive afforestation in the Weihe River Basin, and the reasonable scale of forest and grass planting needs to be determined urgently.

REFERENCES

Budyko, M.I., 1974. Climate and Life. Academic, New York.

Deng, W., Song, J., Sun, H., et al., 2020. Isolating of climate and land surface contribution to Basin runoff variability: A case study from the Weihe River Basin, China. Ecological engineering, 153, 105904.

Duethmann, D., Bolch, T., Farinotti, D., et al., 2015. Attribution of streamflow trends in snow and glacier melt-dominated catchments of the Tarim River, Central Asia. Water resources research, 51(6), 4727–4750.

Eregno, F.E., Xu, C.Y. and Kitterød, N.O., 2013. Modeling hydrological impacts of climate change in different climatic zones. International journal of climate change strategies and management, 5(3), 344–365.

Feng, X., Fu, B., Piao. S., et al., 2016. Revegetation in China's Loess Plateau is approaching sustainable water resource limits. Nature climate change, 6(11), 1019–1022.

Huo, J., Liu, C., Yu, X., et al., 2021. Effects of watershed char and climate variables on annual runoff in different climatic zones in China. Science of the total environment, 754, 142157.

Ma, Z. and Fu, C., 2003. Interannual characteristics of the surface hydrological variables over the arid and semi-arid areas of northern China. Global and planetary change, 37(3–4), 189–200.

Moriasi, D.N., Arnold, J.G., Van Liew, M.W., et al., 2007. Model evaluation guidelines for systematic quantification of accuracy in watershed simulations. Transactions of the ASABE, 50, 885–900.

Pielke, R.A., 2005. Land use and climate change. Science, 310, 1625–1626.

Setti, S., Maheswaran, R., Radha, D., et al., 2020. Attribution of hydrologic changes in a tropical river basin to rainfall variability and land-use change: Case study from India. Journal of hydrologic engineering, 25(8), 05020015.

Shi, Y., Shen, Y., Kang, E., et al., 2007. Recent and future climate change in northwest China. Climatic change, 80, 379–393.

van Roosmalen, L., Christensen, B.S. and Sonnenborg, T.O., 2007. Regional differences in climate change impacts on groundwater and stream discharge in Denmark. Vadose zone journal, 6(3), 554–571.

Wang, S., Fu, B., Piao, S., et al., 2016. Reduced sediment transport in the Yellow River due to anthropogenic changes. Nature geoscience, 9(1), 38–41.

Xu, X., Yang, H., Yang, D., et al., 2013. Assessing the impacts of climate variability and human activities on annual runoff in the Luan River basin, China. Hydrology research, 44(5), 940–952.

Yeşilırmak, E. and Atatanır, L., 2016. Spatiotemporal variability of precipitation concentration in western Turkey. Natural hazards, 81(1), 687–704.

Yin, J., He, F., Xiong, Y.J., et al., 2017. Effects of land use/land cover and climate changes on surface runoff in a semi-humid and semi-arid transition zone in northwest China. Hydrology and earth system sciences, 21(1), 183–196.

Zhang, L. and Qian, Y., 2003. Annual distribution features of precipitation in China and their interannual variations. Acta meteorological sinica, 17(2), 146–163.

Zhang, X., Dong, Q., Cheng, L., et al., 2019. A Budyko-based framework for quantifying the impacts of aridity index and other factors on annual runoff. Journal of hydrology, 579, 124224.

Zhao, Q., Ye, B., Ding, Y., et al., 2013. Coupling a glacier melt model to the variable infiltration capacity (VIC) model for hydrological modeling in North-Western China. Environmental earth sciences, 68(1), 87–101.

CHAPTER 9

Conclusions

9.1 DEVELOPMENT OF THE CHINA WATER ASSESSMENT MODEL (CWAM)

The expansion of hydrological and water resources modeling from the small-/medium- to large-scale regions, means not only wider simulation domains but also greater complexity of geographical and climatic conditions. The traditional distributed hydrological models with physical basis are generally applied on an intact catchment. These models have limitations in describing the characteristics of different climatic and hydrological conditions, geological structures, and their impacts on water yield in large-scale regions. Therefore, a more complex model structure and parameters are required to simulate the behaviors of water yield in a large-scale region like China. Based on the WEP model, a national-scale dynamic water resources assessment model in China, called CWAM, was developed.

Compared with the original WEP model, the main improvements in the CWAM reflected in: ①A sub-basin division method for large-scale complex terrain area was proposed. It balances the requirements of accuracy and computational efficiency of model simulation; ②The relationships of meteorological and vegetation data with elevation were quantified, and their impacts on data interpolation were taken into account; ③The impact of special vadose zone structures, including karst development, swelling soil, and frozen soil, on precipitation-runoff process, was characterized, and the related parameters on soil moisture movement were optimized accordingly.

Simultaneously, the basic hydrological database and common hydrological parameters across the country have been established, which is beneficial for hydrological modeling in ungauged areas. Continuous simulations of various natural hydrological processes were conducted for 62 years from 1956 to 2017. By comparing simulated and statistical monthly streamflow at 203 hydrological stations, the efficacy of the model was verified. For the validation period of 1981–2000, the Nash-Sutcliff Efficiency coefficient (NSE) at 80% of the stations was larger than 0.7, and the absolute value of relative error (RE) was less than 10% at 95% of the stations. The results highlighted the benefit of incorporating new mechanisms on the special vadose zone water movement and accounting for the impact of elevation change on meteorological and vegetation variables.

DOI: 10.1201/9781003646648-9

Because the hydrological phenomenon in the large-scale region is complicated, revealing it precisely is a process that requires continuous improvement. First, the meteorological data are the critical input for hydrological modeling, and the high-quality meteorological data are always desired by hydrologists. Through fusion of remote sensing products, reanalysis datasets, and station observations, several high-resolution meteorological forcing datasets have been produced. These datasets can significantly improve the accuracy of meteorological data in large-scale regions, especially in lack-data basins. Although the dataset is difficult to apply directly to the long-term modeling of this study, it can provide a useful reference for meteorological data input. Besides, the precipitation gradient and lapse rate of air temperature vary on diurnal and seasonal time scales. Therefore, it is necessary to analyze the temporal and spatial variability of meteorological data with elevation in detail when interpolating the data from stations to computation units. Second, in the current study, the inconsistency between the boundaries of surface and aquifer failed to be considered. Considering that this study focused on a large-scale region, the streamflow error caused by this inconsistency was relatively small. In the future research, the inconsistency between the boundaries of surface and aquifer needs to be strengthened to improve the accuracy of the CWAM in localized areas. Third, the current study focused on the simulation of natural hydrological processes for various land cover conditions but did not consider the effects of reservoir, water withdrawal, irrigation, and wastewater reuse. Therefore, the next research plans to couple more social activities into the CWAM to improve the model's applicability in areas affected by strong human activities.

9.2 APPLICATION OF THE CWAM

Through the CWAM, the spatiotemporal patterns of hydrological and water resource variables in China and its WRRs were revealed. The four key hydrological variables (P, R, Inf, and ET_a) showed spatial patterns of gradually increasing from northwest to southeast. During the 62 years (1956–2017), these key hydrological variables in different regions showed complex tendencies. The variables in the northwestern region had a concentrated and large-scale increase trend, which was attributed to the increase in water availability caused by climate change. In the southeast coastal area, the rise in P contributed to the increase in the R and Inf, while ET_a dropped unexpectedly due to reduced vegetation area. Human activities significantly affected the key hydrological variables in local regions but were weaker in scope and intensity than climate change.

We evaluated the amount of water resources during the period 1956–2017 in China based on the output variables of the CWAM. The results showed that the long-term mean value of total water resources in China was $2.81×10^{11}$ m^3/year, of which the values of surface water resources and groundwater resources were $2.63×10^{11}$m^3/year and $0.82×10^{11}$m^3/year, respectively. Moreover, the average overlap of water resources between surface water resources and groundwater resources was $0.64×10^{11}$m^3/year. The RegCM4 Regional Climate Model (RCM4) Medium Future Emission Scenario (RCP4.5) for greenhouse gases and aerosols was selected as inputs to drive the national model, and then the changes of water resources in China for the period 2021–2050 were projected. At the national level, there is a slight increase in future water resources relative to the baseline period

(1956–2017), with a relative rate of change of about 2.8%. Among 10 Class I WRRs, there was a decreasing trend in the water resources of YZRB, PRB, and SWRB. In contrast, the northern river basins showed an increasing trend, with LRB and NWRB demonstrating a notable increase of over 10%, followed by HRB. The remaining Class I WRRs had small changes in water resources. According to the inter-annual variation of water resources in Class I WRRs, it can be found that the northern regions are facing increasing pressure of water shortage. The HRB and YRB, which are characterized by high population density and active economic activity, have experienced a consistent decline in water resources. The recent increase in water resources at NWRB was potentially unsustainable. Reduced water resources in SRB threatened local ecology and food security. In the future, several engineering and policy measures should be taken to optimize the trade-offs and synergies relationship between socio-economic development and water resources.

Furthermore, regional differences in changes of streamflow and its composition among 21 representative river basins, located in nine climatic zones and four geomorphic regions across China, were revealed. The results indicated that the warming and drying trend in the basins in northern China appears to be ongoing. On the contrary, the Heihe River basin and the cold northeast regions were under climatic warming and wetting. In summary, climatic conditions exerted a substantial influence on the magnitude and annual distribution of streamflow. Geomorphic structure, on the other hand, primarily influenced the composition of streamflow. The basins located in the Warm Temperate Zone, which are characterized by both intense human activities and a fragile ecosystem, had a substantial decline in natural streamflow. The frozen soil and karst structures promoted groundwater–river interaction and increased the proportion of river baseflow in streamflow. The contributions of changes of climate and land use to streamflow in each basin were quantified comparatively. Climate change was the predominant factor contributing to the variation in streamflow for most basins (86% of the total), with the contribution rate ranging from 51.4% to 95.7%. Affected by land use change, the streamflow generally decreased in the Warm Temperate Zone but mainly increased in the Alpine Sub-Frigid Zone and Median Temperate Zone. The Grain to Green Programme contributed 55.5% of streamflow reduction in the Weihe River basin. Conversely, the land use change, characterized by drastic reduction of forest, grassland, and wetland, dominated streamflow increase in the Huaihe, and Hulan River basins. The contribution rates for these changes were 65.3% and 66.1%, respectively.

For Product Safety Concerns and Information please contact our EU
representative GPSR@taylorandfrancis.com
Taylor & Francis Verlag GmbH, Kaufingerstraße 24, 80331 München, Germany